Engineering and Commercial Functions in Business

Engineering and Commercial Functions in Business

W. Bolton

Newnes
An imprint of Butterworth-Heinemann Ltd
Linacre House, Jordan Hill, Oxford OX2 8DP

A member of the Reed Elsevier plc group

OXFORD LONDON BOSTON
MUNICH NEW DELHI SINGAPORE SYDNEY
TOKYO TORONTO WELLINGTON

First published 1994

British Library Cataloguing in Publication Data
Bolton, W.
1 Engineering and Commercial Functions in Business
I. Title
658.002462

ISBN 0 7506 2197 4

Library of Congress Cataloging in Publication Data
A catalogue record for this book is available from the Library of Congress

Printed and bound in Great Britain

Contents

Preface

This book has been written to provide a comprehensive coverage of the mandatory unit in Engineering and Commercial Functions of Business for the Advanced GNVQ in Engineering. It aims to introduce engineering students to the functions of business relevant to engineers. No prior knowledge is assumed but it is expected that it will be used in an engineering context. As such it will also be relevant to engineering students on other courses.

The aims of the chapters and their relationship to the elements of the GNVQ unit are:

	Aim	*GNVQ Element*
Chapter 1	Identify business functions, types of business, and the activities of engineers in organizations.	1.1
Chapter 2	Describe organizational structures and factors affecting them.	1.1
Chapter 3	Identifies functions within business and the interfaces between commercial and engineering functions.	1.1 1.2
Chapter 4	Identifies financial factors that influence decisions and explores them using break-even analysis.	1.2
Chapter 5	Identifies the cost elements and investigates absorption and marginal costing techniques.	1.3
Chapter 6	Describes how budgets are determined and used to control costs.	1.3
Chapter 7	Explains the reasons for inventory control and basic techniques that can be used.	1.3
Chapter 8	Explains the role of forecasting and basic techniques that can be used.	1.3

There are many worked examples in the text. In addition, at the ends of each chapter there are a large number of revision questions, multiple-choice questions, assignments and case studies. Answers are given for all the multiple-choice questions and guides indicated as to the answers for the revision questions.

W. Bolton

1 Organizations

1.1 Organizations

The early chapters in this book are concerned with giving an overview of the engineering and commercial functions in business and later chapters with the financial factors involved in decision-making and the costing of products in business. Chapters 1, 2 and 3 are about organizations. This chapter is an introduction to the basic terminology used with organizations, with chapter 2 being a consideration of the structure of organizations and chapter 3 the functions occurring in organizations.

The terms business and organization occur frequently in this book and it is worthwhile clarifying their meaning. The term *business* is used for any person or group of people who sell goods or services. Thus a business might be an individual who works alone and cleans windows. It also could be a large company employing thousands of people and who make cars. The term *organization* is used for a group of people working together over a period of time to achieve a common objective. Thus a business which employs a number of people to make cars is an organization, the group of people having the common objective of making cars. A government department is also an organization, the group of people working together to carry out a government service to the people.

1.1.1 Functions

Within a business, whether it be a small one or a large one, there will be a number of functions that have to be carried out. These are likely to include (note that sometimes some of the functions are grouped under a single heading and sometimes split into smaller functions, functions are discussed in more detail in chapter 3):

1 *Finance* This involves deciding on how much money is required, where it is to come from, how it is to be used, etc.
2 *Purchasing* This function involves the purchasing of the materials, components and energy required for production purposes and the maintenance of the plant and buildings.
3 *Marketing* This involves deciding on the product, the price, promotion of the product, etc.
4 *Sales* This is likely to include dealing directly with the customers or consumer outlets, obtaining orders, advising customers, etc.
5 *Distribution* This involves providing the means by which customers obtain the product, packaging and shipping.
6 *Research and product development* This involves research into new materials and processes and the development of new production methods and products.
7 *Design* This involves the design, and redesign, of the product/service to meet market requirements.

8 *Production/manufacturing* This function involves deciding on how, where, what with, when the items/services should be produced, etc.
9 *Maintenance* This involves the maintenance of the buildings, plant and machinery.
10 *Quality control* This involves ensuring that products and services are to the required quality and can involve the testing and checking of incoming materials, and inspection of the product.
11 *Planning and control* This involves the planning and control of the production function to ensure that it is carried out efficiently and to schedule.
12 *Product support* This involves providing maintenance for products.
13 *Stores and warehousing* This involves the storage of the goods and materials delivered to the business by suppliers, issuing them as required for production, maintenance and administration, and the storage of the finished products for issue to customers as required.
14 *Personnel* This function involves deciding on who to employ, how to motivate them, Health and Safety requirements, conditions of employment, etc.

A single individual operating as a business has to do all the functions required for that business. Thus the person working for him/herself as a window cleaner has to market the product, i.e. obtain the customers for his/her services, produce the service, i.e. do the window cleaning, deal with the finances of the business, i.e. decide how much to charge and on what items money will be spent, keep the accounts, etc., maintain the equipment, office and company van, etc. In a larger business different individuals or groups of individuals will deal with the different functions, each group specializing in a particular function or group of functions. Within any one group there is also likely to be further specialization. Thus, for example, in a production group some employees might do one type of production operation and some another. The term *division of labour* is used when there is a need to subdivide the activities involved in running a business.

Specialization/division of labour occurs in organizations and coordination of the different areas of specialization is required if an organization is to meet a common objective. Organizations are structured so that there is a framework within which individuals can carry out specified duties and which serves as a basis for implementing the procedures required to manage and control the business. Chapter 2 discusses such structuring.

Example

Identify the functions which are likely to occur within the following case study of a small company:

XYZ is a company making lawn mowers. It has a workforce of about 100. The sales staff, sales administration and a small production line account for most of the staff. Salesmen/women visit garden centres and shops around the country to obtain orders. The demand is seasonal with garden centres and shops stocking up in Spring. The orders are met by the company delivering, by their own vans, the mowers from stocks held in the company warehouse. In order to have a steady workload over the year, stocks are built up during the winter months in anticipation of sales in the Spring. The design of the mowers is not changed very often but, because of competitors, there is great pressure on the selling price and quality. Thus the company has to continually monitor its production methods to ensure they are the most efficient and also maintain strict quality control. The company buy-in some of the components and raw materials, doing some of the manufacturing in the company. It has a small production line to assemble the mowers.

If we analyse the case then the functions that are likely to occur with the business are:

Case	Function
XYZ is a company making lawn mowers. It has a *workforce of about 100*. The sales staff, sales adminstration and a small production line	Personnel
account for most of the staff. *Salesmen/women* visit garden centres and shops around the country to obtain orders. The demand is seasonal with garden centres and shops stocking up in Spring. The orders are met by the company	Marketing, sales
delivering, by their own vans, the mowers from	Distribution
stocks held in the company *warehouse*. In order	Stores/warehouse
to have a steady workload over the year, stocks are *built up* during the winter months in	Production planning and control
anticipation of sales in the Spring. The design of the mowers is not changed very often but, because of competitors, there is great pressure on the selling price and quality. Thus the company has to continually	Finance
monitor its production methods to ensure they are the most efficient	Product development
and also maintain strict *quality control*. The	Quality control
company *buy-in some components and raw*	Purchasing
materials, doing some *manufacturing* in the company. It has a small production line to assemble the mowers.	Production

1.2 Inputs and outputs

We can consider any business in terms of what its inputs and outputs are. Inputs with a manufacturing business are typically materials, energy, labour and money and outputs are finished products and waste (figure 1.1). A business making cars can be considered as one which has an input of materials, energy, labour and money and an output of cars and scrap. A college can be considered to be an organization with an input of untrained students, labour and money and an output of trained and failed students. A bank can be considered to be an organization with an input of money and labour and an output of money and financial services.

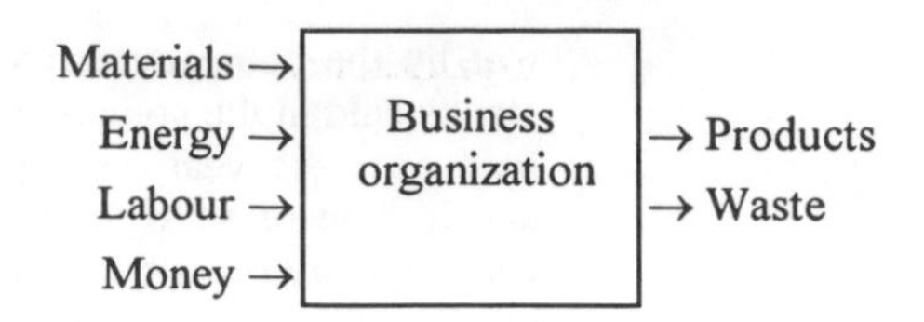

Figure 1.1 A manufacturing business

Production, in one sense, can be considered to be the conversion or transformation of physical inputs into physically different outputs. However, we can regard it in a broader sense and include the provision of services as also being a production activity. Thus we can regard all business organizations as, in a sense, being production organizations. All have inputs, which they transform into their outputs.

Example

What are the main inputs and outputs for a car repair/maintenance business?

The main inputs might be considered to be cars needing repair or maintenance, materials, labour, money, electrical power, etc. The main outputs might be considered to be repaired or serviced cars and waste materials.

Example

What are the main inputs and outputs for a company building houses?

The main inputs might be considered to be materials, labour and money and the main outputs finished houses.

1.2.1 Different types of business

There are many different types of business. They are frequently classified by the type of the production activity they are engaged in. The terms primary, secondary and tertiary are used for the three levels in the classification, these being defined as:

1 *Primary organizations* These are organizations concerned with the first stage of production, having raw materials as their input and output of materials in a more useful form for other production organizations. Examples of the types of activities of such organizations are mining, agriculture, fishing and forestry.
2 *Secondary organizations* These take the end product of the primary organizations, or other secondary producers, and further transform them into new products. They make things and are generally referred to as the *manufacturing* or *construction* industries. Examples of such

organizations are businesses which make cars, washing machines, aircraft or build houses or factories.

3 *Tertiary organizations* These offer services to people and other businesses. Such services include retailing, insurance, banking, teaching, medical services, repair services, transport, etc.

Example

With what type of production activity could (a) a motor manufacturer, (b) a coal mining company, (c) a shop, be associated?

(a) The motor manufacturer has inputs of materials such as sheet steel, fittings, paint, etc., energy, labour and money and an output of cars for sale to the public. The end products of primary and other secondary organizations are transformed into new products and so the motor manufacturer is a secondary production organization.
(b) The coal mining company is engaged in the removal of coal from the earth, processing it to yield fuel which is graded by quality and size, and then selling it to others to use. The company can be classified as a primary organization.
(c) A shop has inputs of manufactured goods, labour and money and an output of customers with the goods. The end products of secondary organizations are provided as a service to customers. The shop can be classified as a tertiary organization.

1.2.2 The chain of production

Businesses tend to be linked in a chain of production which typically stretches from the raw materials being mined and refined by a tertiary business, the refined products being bought and used by secondary businesses and tertiary businesses providing services to support both the primary and secondary businesses (figure 1.2).

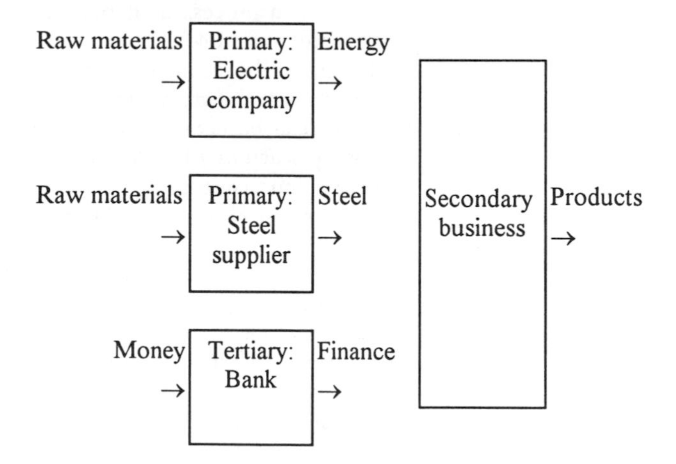

Figure 1.2 A simple chain of production

Thus, some businesses sell the products and services they produce to other businesses. For example, a motor manufacturer will buy-in such items as steel sheets, steel bars, and other materials in a form that has been prepared by primary businesses. In addition they are likely to buy-in components such as lamp mountings, electrical meters, etc. from secondary businesses. They also are likely to buy-in services such as retailers, banking, insurance, etc. from tertiary businesses. It has been suggested that half of those employed in non-manufacturing organizations depend for their jobs on the links with the manufacturing sector.

1.3 Engineering organizations

In the engineering industry in Britain, over 80% of the businesses have less than 100 employees. However, because the large organizations, though far fewer than the small ones, are very large, most of those who work in the engineering industry are employed by large organizations. The large businesses were however, once small. Why do businesses expand? The major reason is to increase profits. Businesses expand by what is termed horizontal integration, vertical integration or lateral integration.

The term *horizontal integration* is used to describe the expansion of an organization when it increases the scale of its operations while continuing to specialize in the same range of products. The increase in output of the same products is expected to give a less than proportionate increase in costs, i.e. as outputs increase there is a reduction in unit costs. The organization is said to be benefiting from *economies of scale*. Economies of scale can arise from:

1. The larger volume output allowing a greater division of labour, i.e. employees can become more specialist and concerned with a narrower range of skills and hence more proficient in the narrower range. This specialization may mean that unskilled labour can be used for some jobs where previously a skilled person had to be used because of the range of skills required. Hence the labour costs per unit can be reduced.
2. The larger volume output can lead to a greater mechanization of the production process as it becomes more economic to use machines if they are used a lot.
3. There can be economies in purchasing materials and items because larger quantities are required.
4. There can be economies in marketing. For example, the distribution costs per unit can decrease because the distribution costs may not rise in proportion to quantity.
5. Finance may be easier to obtain and on better terms.

The term *vertical integration* is used when an organization expands to take in more of the elements involves in its chain of production, e.g. engaging in both primary and secondary stages of production. For example, a steel company might not only extract the raw materials, produce iron and steel, shape the steel into semi-finished products such as sheet and bar, but also produce some finished products, e.g. railway lines. This may lead to economies through giving greater efficiency and also increase the power of the organization over the market.

The term *lateral integration* is used when an organization expands by diversifying its activities, i.e. increasing the range of products. Such diversification can reduce reliance upon a single product and so reduce the risk factor. For example, a maker of aircraft engines might diversify into the production of electric power generators.

However, the larger an organization the greater the management problems. The rise in the costs required to administer and control the organization might outweigh the advantages gained by economies of scale. There is also the point that increased division of labour can lead to employees performing such narrow tasks at such a frequent rate that they become bored and lose motivation with the result that a deterioration in labour relations occurs.

1.3.1 Employment in the engineering industry

The workforce employed in the engineering industry has been classified (EITB Research Report 9, The technician in Engineering: Part 1 Patterns of technican employment) as being made up of:

Managers	5.4 %
Scientists and technologists	3.0 %
Technicians	8.3 %
Administrative and professional staff	6.0 %
Supervisors	5.1 %
Craftsmen	19.1 %
Operators	33.8 %
Clerical staff	11.1 %
Others	8.2 %

The employment of engineering technicans in the engineering industry was in a range of different jobs. In the survey of the industry, in general it was found that among the engineering technicans about:

1. 50% were working in production as production or planning engineers/technicians, installation or commissioning engineers, software engineers, maintenance engineers, test engineers/technicians, project engineers, quality engineers, inspectors, etc.,
2. 20% were working in the research, design and development area in such capacities as design draughtsmen/women, detail draughtsmen/ women, design engineers, development engineers, etc.,
3. 20% were working in areas involving customers or suppliers as service engineers for customer products servicing, contract engineers, buyers, estimators, technical sales, etc.,
4. 10% were working in central services such as work study engineers, technical writers/authors, technical illustrators, service engineers for internal company servicing, etc.

Problems

Revision questions

1 Explain what is meant by the terms (a) business, (b) organization.

2 State the activities implied by the business functions of (a) marketing, (b) production, (c) personnel.

3 Explain the terms (a) primary, (b) secondary, (c) tertiary when applied to organizations and give an example of each.

4 Explain what is meant by horizontal, vertical and lateral integration when applied to businesses.

5 Explain what is meant by economies of scale.

6 List five types of jobs that might be undertaken by engineers in an organization.

Multiple choice questions
For problems 7 to 16, select from the answer options A, B, C or D the one correct answer.

7 The following describe activities:

(i) Inspecting products
(ii) Planning tool changes and settings
(iii) Ordering materials

Which option BEST describes those activities that might be involved in the quality control function?

A (i)
B (i) and (ii)
C (ii) and (iii)
D (i), (ii) and (iii)

8 Which one of the following businesses can be classified as a secondary business?

A A college
B A garage
C A shop
D A furniture maker

Questions 9-11 relate to the following information.

A business carries out the following activities:

A Negotiating a loan to pay for new equipment

B Advertising a new product
C Dealing with a dispute concerning pay
D Operating a lathe

Select from the above list the activity with which the following functions within a business will be most concerned:

9 Marketing

10 Production

11 Finance

12 Decide whether each of these statements is True (T) or False (F).

(i) A primary organization has an input of raw materials
(ii) A secondary organization provides a service to other organizations and people

Which option BEST describes the two statements?

A (i) T (ii) T
B (i) T (ii) F
C (i) F (ii) T
D (i) F (ii) F

13 Decide whether each of these statements is True (T) or False (F).

The activities of engineers in a manufacturing organization can include:
(i) production inspection
(ii) servicing products
(iii) technical sales

Which option BEST describes the two statements?

A (i) T, (ii) F, (iii) F
B (i) T, (ii) T, (iii) F
C (i) F, (ii) T, (iii) F
D (i) T, (ii) T, (iii) T

14 Functions that might be present in a business are:
(i) Production
(ii) Quality control
(iii) Purchasing

An electronic company makes CD players. It buys in the components and assembles them to obtain the finished products which it puts its brand name on. Which of the functions are most likely to be present in the electronic company?

A (i)
B (i) and (ii)
C (ii) and (iii)
D (i), (ii) and (iii)

15 Businesses can be classified as primary, secondary or tertiary. An example of a primary business is:

A Construction
B Insurance
C Agriculture
D Repair

16 Businesses can be classified as primary, secondary or tertiary. An example of a tertiary business is:

A Instrument maker
B Shipbuilder
C Furniture maker
D Hospital

Assignments

17 List ten organizations that are in your locality and classify them as primary, secondary or tertiary.

18 For a local business, list the likely other businesses that it will use.

19 List three different types of local businesses which you consider will have engineers employed, giving three examples in each case of the types of jobs they could be doing.

20 Consider some item you have recently purchased. Identify a possible chain of production which might have been involved from raw materials to the finished product.

Case studies

21 John was interested in bicycles and in his spare time, as a favour to friends, bought components and assembled tailor-made bicycles for his friends. The number of friends increased rapidly and after a while John decided to quit his job in a local factory and start his own business producing tailor-made bicycles. The business grew and, after a while, John found he could not cope with all the work. He then took on two workers. John still did some of the assembly work, along with the two workers, but dealt exclusively with the buying-in of the parts and sales to customers. The business grew and John took on more workers. He now found he had insufficient time to do any assembly work and had to leave all that to his workers. He also found that he had to have one of

the workers specialize in dealing with the customers while he concentrated on the financial side and controlling and co-ordinating the activities of the workers.

(a) In what category could John's business be classified?
(b) Why, as the business grew, did John find it necessary for him and his workers to specialize?
(c) Identify the functions that are likely to be present in the business when it just consisted of John and then later when it had grown.
(d) Why do you think John wishes to increase the size of his business?
(e) Explain how this case study is an example of division of labour and list the benefits that might be obtained from such a division. What additional work will result from such a division?
(f) After a period of growth, John makes one of the original assembly workers a superviser for the assembly workers. A month later the new supervisor complains to John that, though he liked being paid more, it really was not as important a job as assembling bicycles, since all he dealt with was the chasing up of items for the assembly workers and dealing with their problems. What type of reply do you think John should give?
(g) What chain of production might be involved in John's business producing tailor-made bicycles?
(h) After a period, John considers the business ought to expand beyond just the single factory producing bicycles. What types of expansion might be considered.

22 As part of the studies for its full-time engineering students, to give them a taste of the world of business, a college plans to set up with the students a small business which will make pocket-size cassette players. The setting up and organization of the business is the responsibility of the students, with guidance from the college lecturers. The students, after being given the assignment, have a meeting at which there is considerable discussion about what should be done, and ends up with no direct conclusion other than that all of them will go away and produce ideas for the next meeting. At the next meeting one of the students produces the plan that they should divide up the tasks required with designated students having responsibility for each task.

(a) List the types of tasks/functions that they will need to consider.
(b) Will the business be a primary, secondary or tertiary business?
(c) Work out a possible chain of production, involving external businesses, with which the college business is likely to be part of.

2 Structure of organizations

2.1 Why have a structure?

This chapter is about the structure of organizations and opens with a discussion of a very simple organization before moving on to consider more complex organizations and their structures.

A person going into business on his or her own account has to carry out all the functions of that business. Thus, suppose we have Ann who is going into business on her own account fitting car exhausts. She has to establish the market that is available for the product, obtain orders, buy all the materials and components required, arrange storage for them, plan the manner in which the exhausts will be fitted, carry out the fitting, obtain payment for the work, pay for the materials purchased, etc. The structure of the business is very simple since it consists of just one person. However, if the business expands and Ann takes on employees then division of labour can be used by Ann so that some of the workforce specialize in, say, fitting exhausts, some in dealing with the finance and payments, some on obtaining orders, etc. With just perhaps one or two employees the co-ordination of the work might be the simple process of informal communications between them, i.e. no formal arrangement about who issues orders but just chatting between the workers with control of the work, and responsibility for the work, resting in the hands of those doing the job. But with more employees direct supervision might be necessary to achieve co-ordination. All the workforce might then operate directly to Ann who assumes a supervisory/managerial role, co-ordinating all their activities. Co-ordination is then achieved by one individual taking responsibility for the work of others.

We can represent this supervisory arrangement by an *organization chart* (figure 2.1). This shows the structure of the organization. The lines joining the employees to Ann represent a direct formal relationship between them. The fact that all the lines emanate from Ann means that she alone has responsibility over the others. Thus any query from an employee must be raised directly with Ann. The reason for the business being structured in this way is because Ann considers that division of labour will result in a more efficient and hence more profitable business and the structure then indicates the way she will achieve co-ordination of the labour.

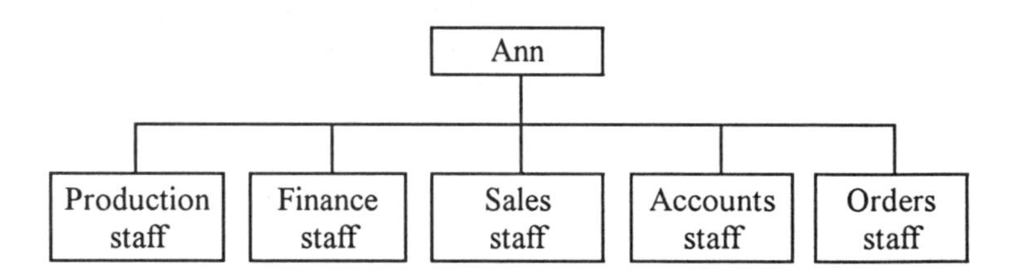

Figure 2.1 The organization chart

In a larger organization a situation similar to that shown in figure 2.1 would rapidly lead to chaos. Consider the chaos that would result in a larger company with, say, a 1000 employees if each employee had to direct his or her queries to just one person. Thus larger organizations have a structure where all the decisions are not taken by one person but authority is delegated to others to make decisions.

To sum up: the term *structure* is used for an organization to mean the ways in which it divides its labour into distinct tasks and then achieves co-ordination among them and can be represented by an *organization chart*. The structure thus indicates the allocation of formal responsibilities and the formal links between employees.

Example

Harry has decided to set up his own business developing computer software for production planning in manufacturing organizations. Initially he plans to start in a small way with a workforce consisting of himself as manager, five programmers, a secretary and an accounts clerk. All the employees are to report directly to Harry. Sketch the organization chart for the business.

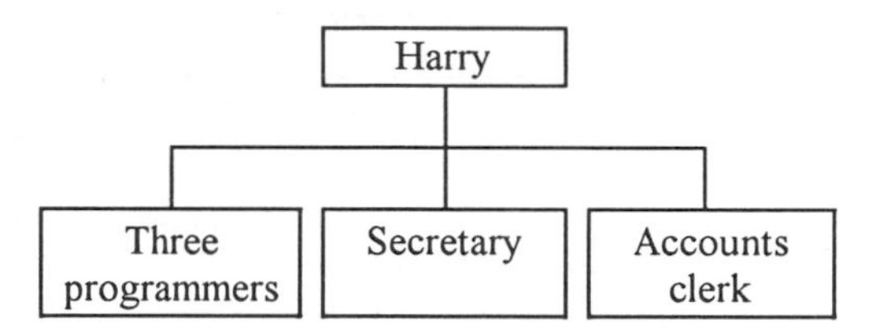

Figure 2.2 Example

2.2 Authority

In discussing organizations and their structure, terms such as authority, chain of authority, line authority, staff authority, functional authority, responsibility and delegation are used. The following are explanations of such terms.

Authority is the right to make decisions, give orders and direct the work of others. However, there can be some *delegation* of authority. This means that if a supervisor delegates some authority to, say, a foreman then he or she will be able to make some decisions and direct work. Delegation involves the subordinate being accountable to the superior for the performance of the tasks assigned to him or her. The subordinate can, however, only be assigned some degree of responsibility, the ultimate responsibility still resides with the superior. Thus while authority can be delegated, respons- ibility cannot be entirely delegated.

The production manager may give orders to the supervisor. The supervisor may give orders to the foremen. The foremen may give orders to the workers. Such a sequence of authority delegation is called a *chain of authority*. The lines in the organization chart show the chains of authority. Figure 2.3. illustrates such a chain of authority, showing how the chain extends from the production manager, down through supervisor A, to foreman A to workers A, B, C, etc. A person is said to have *line authority* if he or she appears within such a chain, i.e. he or she receives orders from above but has delegated authority which enables him or her to give orders to those below him or her in that particular chain.

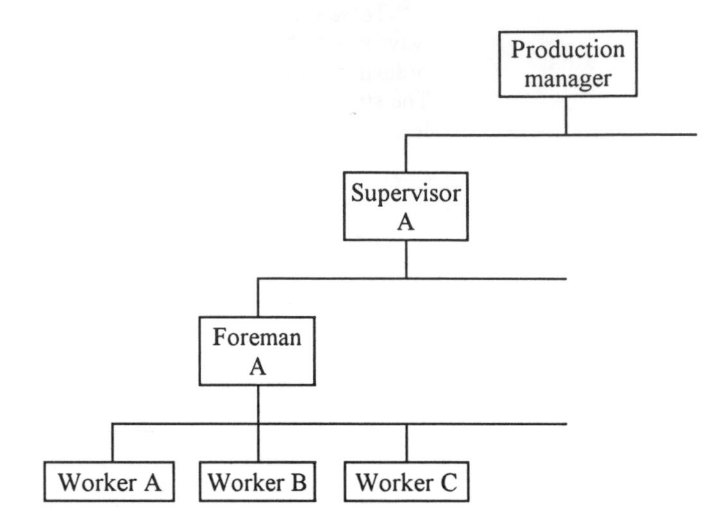

Figure 2.3 A chain of authority which extends from production manager to workers A, B and C

A person is said to have *staff authority* if he or she is not in the chain of authority but is in the organization in an advisory capacity. A personnel manager can be considered to be in an organization in such a capacity, thus having staff authority in relation to the production manger, in that he or she can only advise the production manager. The personnel manager is, however, likely to have some line authority in relation to his or her own staff.

The term *functional authority* is used when an employee is empowered to give orders with respect to just some particular function. Thus the press officer of an organization might have functional authority with regard to dealings with the mass media, no other person being allowed to deal with them without the authority of the press officer.

Many organizations are a mixture of both line and staff authority, with invariably some employees also having functional authority.

Example

Figure 2.4 shows the organization chart for a small manufacturing company. For that organization:

(a) To whom is the production manager accountable?
(b) For whom is the production manager responsible?
(c) What is the span of control of the production manager?

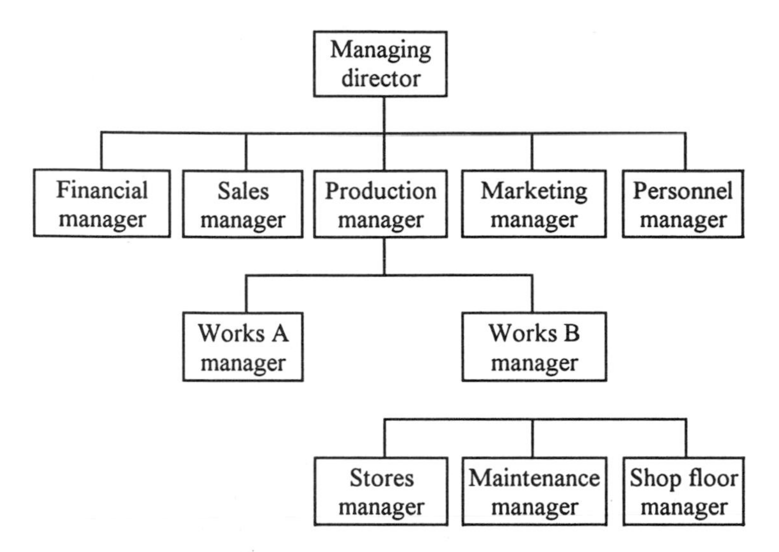

Figure 2.4 Example

The production manager is accountable to the managing director and responsible for works A and works B managers. The span of control of the production manager is just two, being the two works managers.

2.3 Organizational structure

The basic reason for structuring an organization is so that it can be effectively co-ordinated and directed towards the achievement of the objectives of the organization. This is generally done by putting employees in groups. The following are constraints that tend to occur in doing this:

1. Groups should be organized in such a way as to give the best chance of meeting the objectives of the organization. This might mean in some cases organization by function, by locality or by project. By function means organizing in terms of the type of business function carried out, e.g. marketing, production, personnel. By locality means organizing by where the group of workers is located. Thus, for example, an organization might be split into two groups because they are located at two sites. By project means organizing so that all the workers associated with one particular project fall in one group, all the others involved with a different project in a different group. These forms of organization are discussed later in this chapter.
2. No work group should be too large for one person to manage. The number of people any one person can directly control depends on the complexity of the jobs undertaken by the subordinates. The more complex the job the smaller the number of subordinates that can be controlled. The term *span of control* is used for the number of subordinates directly controlled by any one individual. A pyramid structure thus occurs with a *hierarchy of management* as illustrated in figure 2.5.

Chief executive
Senior managers
Each with middle managers
Each with supervisors/junior managers
Each with subordinates/operatives

Figure 2.5 The hierarchy of management

The term *senior management* tends to be used for those who have the authority to determine policy and take decisions to implement that policy without reference to higher authority. *Middle management* tends to be used for those who are given authority to take decisions within the policy laid down by senior management and defined by their specified function. The term *supervisors* is used for those who are given authority to deal with day-to-day problems within the constraints laid down by middle management.

3 No work group should be too small, otherwise the manager of the group would have insufficient work to do and might be better employed in another capacity.
4 The fewer the levels of management the better, i.e. the fewer the links in the chain of authority the better. The more links there are the greater the communication problems. The term *flat organization* is used for one which has few levels, whereas *tall organizations* have many levels. A flat organization thus has a more centralized structure with wide spans of control (figure 2.6(a)). A tall organization is normally more decentralized since there are many levels and narrower spans of control (figure 2.6(b)).

Chief executive
Six managers
Each with many subordinates/operatives

(a)

Chief executive
Two plant managers
Each with four senior managers
Each with five middle managers
Each with five supervisors/junior managers
Each with subordinates/operatives

(b)

Figure 2.6 (a) A flat form of organization, (b) a tall form of organization

2.3.1 Centralization

When all the power for decision making rests at a single point in an organization then the structure is said to be centralized. When managers at the lower levels are given the authority to take decisions that are important to the business then the authority is said to be *decentralized.* The advantages of centralization are that:

1 Decisions can be made quickly.
2 Senior managers, in having an overall view of the business, are more likely to make decisions relevant to the objectives of the business.
3 Senior managers have more experience of making decisions and so are less likely to make bad decisions.

There are, however, disadvantages with centralization, these being:

1 There can be delays in implementing, at a lower level, decisions made at higher levels in the hierarchy.
2 Senior managers may not be able to cope with all the decisions to be made. The sheer number of decisions to be made may result in their losing sight of policy decisions.
3 Senior managers may not be aware of problems occurring at lower levels.
4 It does not allow more staff to participate in the decision making and so they may lack the motivation that comes from being able to exercise their own judgement.
5 Lower managers do not gain experience of decision making.

Centralized structures are characterized by having only a few levels of management since there is little need for middle management if they are given no authority to make decisions. They thus tend to be flat structures. On the other hand, decentralized structures tend to be tall structures with middle management. Thus a centralized structure might look like that shown in figure 2.7 while a decentralized structure looks like that in figure 2.8. With the centralized structure, the financial, sales, production, marketing, personnel, etc. managers all report directly to the managing director. With the decentralized structure, only the plant managers report to the managing director. Each plant manager has his/her own financial, sales, production, marketing, personnel, etc. managers reporting to him/her.

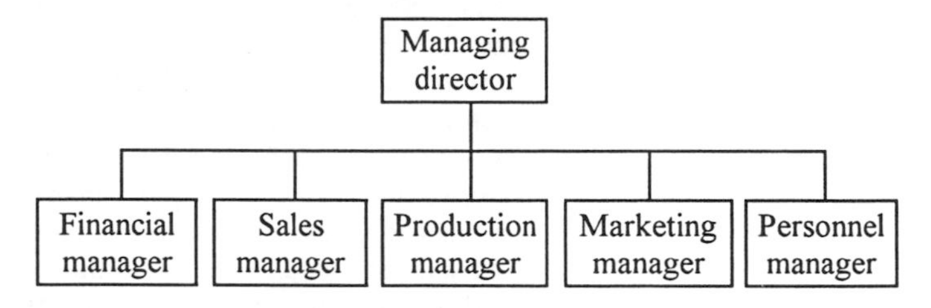

Figure 2.7 Centralized organization

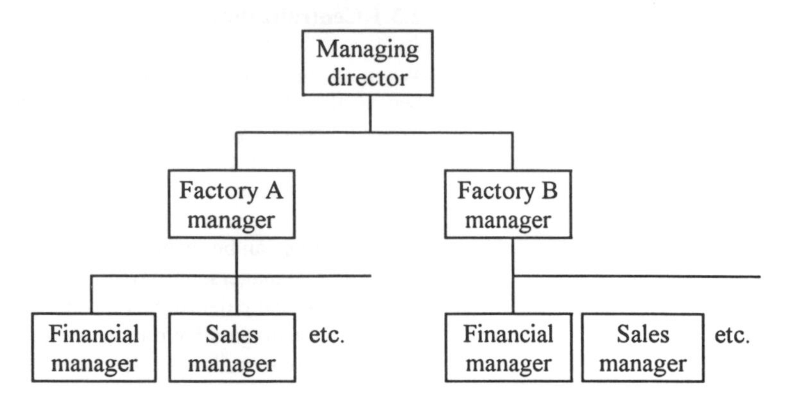

Figure 2.8 Decentralized structure

2.4 Structuring an organization

Basic ways that organizations can be structured are:

1 By function and work process, e.g. production, marketing, sales, etc., and perhaps welding, machining, casting.
2 By location, e.g. the factory in the south and the factory in the north, or by product, e.g. cars and lorries.
3 Matrix.

This section is a consideration of such structures.

Consider the small business described in section 2.1 of Ann who is going into business on her own account fitting car exhausts. Initially she carries out all the functions of the business herself. The structure of the business is very simple since it consists of just one person. However, if the business expands and Ann takes on employees then she has the problem of determining how to divide the work between them. She could perhaps have each one of them doing all the functions and so each essentially behaving as she was when she was working alone. However, it is likely to be a more efficient use of the extra people if specialization occurs so that some of the workforce specialize in, say, fitting exhausts, some in dealing with the finance and payments, some on obtaining orders, etc. This means that she could hire a worker who was very proficient in fitting exhausts but not necessarily good at dealing with account books and keeping track of money. She could hire a person to do sales who was good with talking to potential customers but did not have the skills to fit exhausts. Such specialization is obtained by dividing the organization into *functional areas.* With such a small business the organization structure is likely to be fairly flat. However, if the exhaust centre grew, Ann might need to introduce another layer of management and put a manager in charge of the commercial side, i.e. finance, sales, marketing, and another manager in charge of the production side. She might then group the workers under him/her according to the work process they undertake, e.g. fitting, welding, etc. Such a subdivision is by *work process*.

Division by function involves dividing the organization into functional areas such as marketing, production, distribution, etc. (figure 2.9(a)). Such a structure enables specialist workers to be grouped together. Further subdivision might then be by work process (figure 2.9(b)).

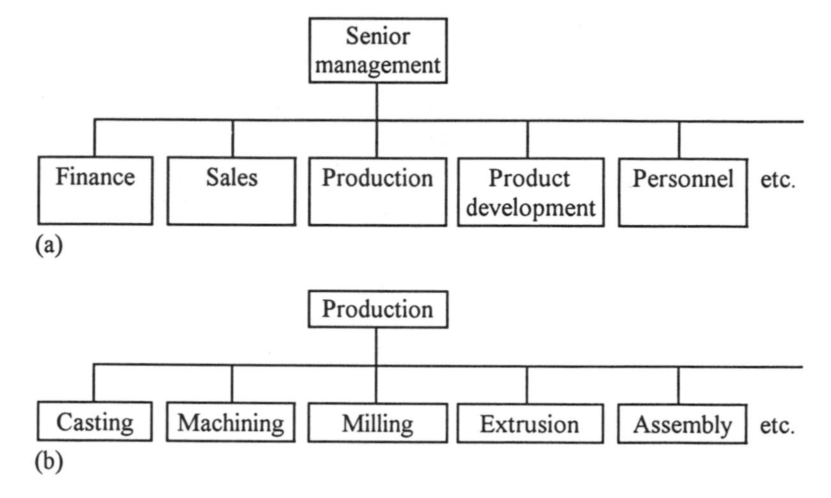

Figure 2.9 Division by (a) function, (b) work process

Now suppose Ann expands her business and sets up exhaust centres in a number of widely dispersed towns. In such a situation she might have a manager at each of the centres with each manager having a functionally- and perhaps process-divided group of staff under him or her (figure 2.10). Ann is likely to require a headquarters where an overall management group can be located. Such a structure for the business has introduced extra layers of management and so is moving from the flat structure which was suitable with just the single exhaust centre to a taller hierarchical structure. We can consider, in this case, there to be a division by *location*.

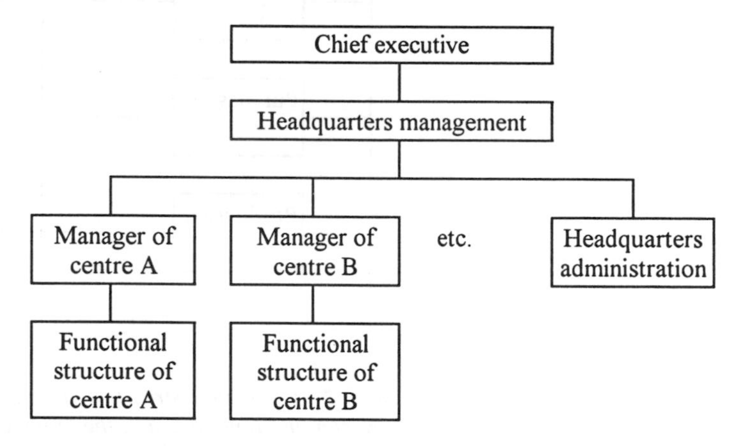

Figure 2.10 Division by location

Division by location involves grouping employees according to the location in which they work. A similar form of division is when employees are grouped according to the product. Thus in a business making both cars and lorries the workforce may be divided into those making cars and those making lorries. With both such structures there can be structures by function and work process at each location or for each product.

Now suppose Ann obtains a contract to maintain the exhaust systems of the fleet of cars owned by a car hire company. She might feel that the most efficient way of dealing with this is to put one person, a team leader, in charge of the contract. That person would, for just this contract, have staff from a number of functional areas reporting to him or her. The employees working in the functional departments of the business remain accountable to their functional manager, but they also report to the contract team leader. Such a structure is termed a *matrix structure*.

Matrix structures are built around individual projects or products and have team leaders co-ordinating the resources drawn from the various functional areas. Figure 2.11 illustrates the form of such a structure.

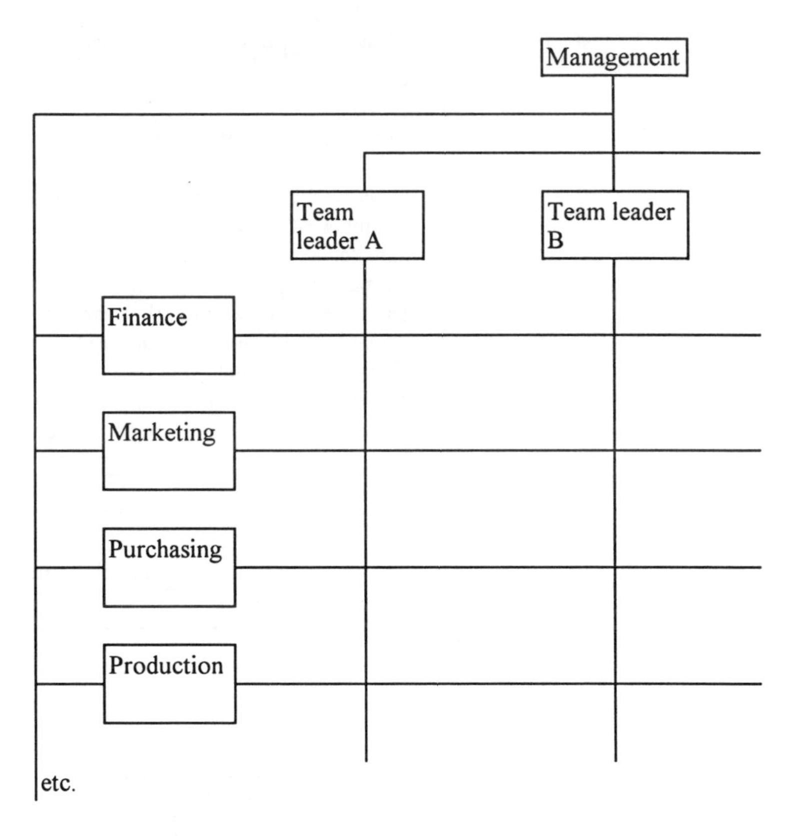

Figure 2.11 A matrix structure. The employees at the intersections of the lines from the functions and the team leaders report to both the functional heads and the team leaders.

2.4.1 Examples of structures

The following figure (2.12) is an illustration of the organizational structure of a medium sized manufacturing business. It could be a secondary business producing, say, plastic mouldings for use in other secondary businesses, e.g. casings for telephones. The business has two main functions: production and administration. Production covers the manufacturing of the products, product development, engineering concerned with maintenance of the plant, stores, and quality control. Administration covers sales, purchasing, accounts and personnel. Each of these functional areas has staff. In the case of the sales area there may be a sales representative for each area of the country. With stores there may be three separate stores, one for general administrative materials, one for incoming materials and one for finished products. With the finished products store there may be an associated distribution section. In the case of the manufacturing area the subdivision is indicated with three shift supervisors, each in charge of a number of operators. This subdivision is a grouping of employees according to the time at which they are working.

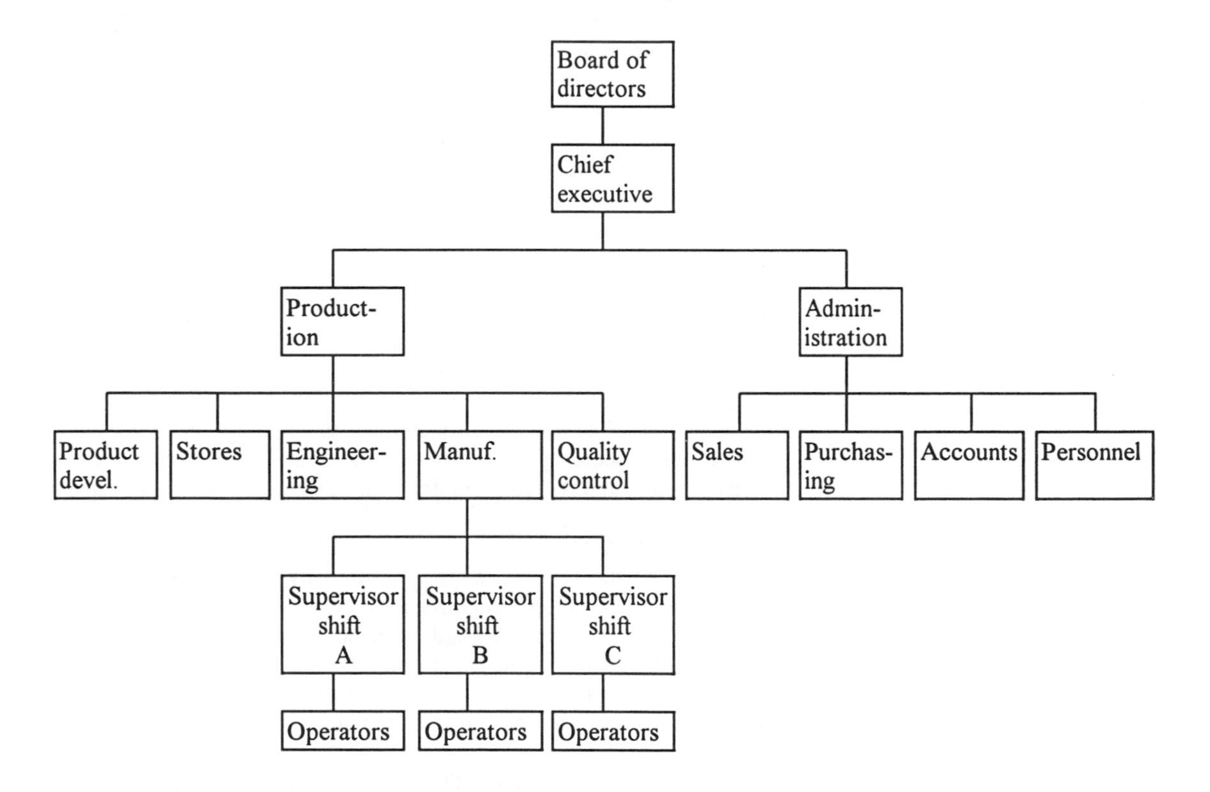

Figure 2.12 An example of the type of organizational structure of a medium sized manufacturing business

The above represents a common type of structure for a medium sized manufacturing business, namely a basically functional structure with few layers of management. But would the structure be different if the business had been manufacturing different products or perhaps providing services? The following section gives an indication of possible answers to such questions.

2.5 Types of production

The type of production operations carried out by a business can determine the type of organizational structure that the business uses. Production operations can be classified into a number of categories, according to the degree of repetitiveness involved. They can be classified as:

1 *One-off production* This is when the production is concerned with producing one, or a very small number, of unique items. The whole of the product run is considered as one operation and work completed on it before another one is started. This is commonly referred to as *jobbing*. It is usually a make-to-order system.
2 *Batch production* This is when larger volumes of a product are required. They may be tackled by taking a batch of items through each stage of the production in turn. Thus an entire batch may be cast before the entire batch goes on to be machined. This system is used for making items against orders received or making items for stock, orders then being met by drawing items from this stock.
3 *Flow production* This involves the use of an assembly line. The term *mass production* is frequently used for this form of production. This has workers, plant and tools positioned at each point along the line and geared to the performance of single tasks at each point, such tasks being constantly repeated as the product proceeds along the line. The car assembly line is an example of such a form of production. Such a form of production can only be used when there is a regular and consistent demand for a product. Items are made for stock. Thus when you order a car it is not specifically made for you but one is taken from stock.
4 *Process* This form of production involves materials being involved in a continuous process. An example of such a form of production is the manufacture of chemicals or materials, where there is a continuous mixing and processing of the raw ingredients and consequently a continuous production of the finished item. This involves items being made for stock.

The form of the production process used by a business can have an effect on the organization structure. J. Woodward carried out a survey between 1953 and 1957 of one hundred manufacturing businesses in Essex, England (J. Woodward, Management and Technology, HMSO 1958, p4). When the businesses were grouped according to the production system used, the following are some of the results obtained:

1 The number of levels of authority increased from an average of about three with one-off or small batch production businesses to four or five with large batch and flow production and six with flow production.

2 The span of control by supervisors of operatives was an average of about 20 to 30 for one-off or small batch production to 40 to 50 for large batch or mass production and 10 to 20 for flow production.
3 The ratio of managers and supervisory staff to total personnel increases as the system of production moves from one-off or small batch, to large batch or mass production and to flow production. For a company with between 400 and 500 employees, with one-off production there was 1 manager/supervisor to 22 other staff, for mass production 1 to 15, and for flow production 1 to 8.

Example

On the basis of the Woodward results, what type of organizational structure might be expected for a company engaged in (a) small batch production, (b) flow production?

(a) With small batch production, the company would be expected to have about three levels of authority with a comparatively small number of managers/supervisors to other staff. Thus a comparatively flat organization structure would be expected.
(b) With flow production, the company would be expected to have about six levels of authority with a large ratio of managers/ supervisors to other staff. Thus a tall organizational structure would be expected.

2.5.1 Stable and innovative environments

In the early 1960s, T. Burns and G. M. Stalker carried out an investigation of twenty industrial companies in Britain (T. Burns and G. M. Stalker, The Management of Innovation, Tavistock 1966). The investigation was to find out how the nature of the environment within which an organization operated affected the type of organizational structure within the company. They evolved two categories of environment, a stable environment and an innovative environment. A *stable environment* is specified as having the following characteristics:

1 The demand for the product or the service supplied by the organization is stable and can easily be predicted.
2 The competitors for the same market are unchanging and stable.
3 Technological innovation and the development of new products is a gradual process and the changes can be predicted well in advance.
4 Government policies concerning the regulation of the industry and taxation are stable and change little with time.

An *innovative environment* has the characteristics of:

1 The demand for the product or service supplied by the organization can change rapidly.
2 The competitors for the same market can change rapidly.

3 Technological innovation and the development of new products occur at a rapid rate.
4 The policies of the government concerning the regulation of the industry and taxation change rapidly.

In their survey, Burns and Stalker found that a rayon mill was operating in a stable environment. It had an organizational structure to which they gave the term *mechanistic*. Such a structure has the characteristics of:

1 Highly specialized jobs.
2 Functional division of work.
3 Close adherence to a rigid chain of command.
4 Centralization of decision making.
5 Small span of control of individual managers.
6 Close supervision of the workers.
7 Functional types of department prevail.

Electronics firms were, however, found to be operating in an innovative environment. The type of organizational structure differed from that in companies operating in a stable environment. Burns and Stalker termed the organizational structure as *organic*. Organic organizational structures have the characteristics of:

1 The jobs are not clearly defined and the workers have to be adaptable.
2 Division of the work is not by function but by task.
3 Little attention is paid to a rigid chain of command.
4 Few decisions are centralized.
5 Workers exercise self-control and there is no close supervision.
6 There is an emphasis on consultation rather than command.
7 Product types of department prevail rather than functional types.

In the later 1960s, P. R. Lawrence and J. W. Lorch carried out an investigation in large, multi-department organizations. They concluded that different parts of an organization may be operating in different environments. Thus different parts of the organization might be structured in different ways. Thus some parts might have a stable environment and some an innovative environment. Thus some parts of the organization might have a mechanistic structure and some an organic structure.

Example

What type of organizational structure might be expected for a business which is manufacturing a product for which there is a constant steady demand with little change in technology or market competition?

The organization can be classified as operating in a stable environment and thus a mechanistic type of structure is likely. This would indicate a functional organizational structure.

2.6 Activities

The general activities of managers and supervisors can be summarized in a number of ways. Traditionally the activities have been seen as planning, organizing, directing, co-ordinating and controlling. The following list includes these terms and some others that might also be considered to be part of the activities of managers and supervisors.

1 *Planning* This activity involves the setting of objectives and targets, the making of predictions, anticipating problems, and planning for future demands. Thus, for example, a production manager in setting the production budget for the coming year might plan for 1000 items of product X to be produced.
2 *Organizing* This involves ensuring that the right resources are available at the right time so that work may proceed according to plan. Thus, for example, a sales manager might organize his/her staff so that there are sales representatives covering each area of the country.
3 *Directing* This involves guiding and supervising subordinates, motivating them to do the job and maintaining morale. Thus, for example, a marketing manager might supervise his/her staff in their devising of a questionnaire to determine the market demand for a new product.
4 *Co-ordinating* This is the directing of the efforts of subordinates so that their efforts all are directed towards the same end, namely the achievement of the laid down plan. Thus, for example, a production manager might co-ordinate the machining and the assembly so that the workers end up with a correctly machined and assembled product of the right quality and in the required time.
5 *Controlling* This is the checking of how the actual performance of subordinates compares with that planned and making adjustments so that performance does lead to the required achievement of the laid down plan. Thus, for example, a sales manager might check during the year that the sales representatives are obtaining the orders that he/she budgeted for in the budget at the beginning of the year.
6 *Deploying/staffing* This involves considering the numbers of staff required and their skill requirements. Thus, for example, a production manager might have to consider how many staff and what skills they require for a new production line.
7 *Liaising* This involves a liaison role dealing with others in the organizations and outside the organization. Thus, for example, the purchasing manager has to liaise with the production manager to find out what materials are required and when, with the finance manager to ensure that there is enough money to pay for the materials, with stores to advise them that orders have been placed and that materials might be expected, with outside suppliers to obtain quotations for materials, etc.
8 *Informing* This involves the monitoring and dissemination of information obtained from higher up in the organization to subordinates and from subordinates to higher up in the organization. Thus, for example, the production manager needs to inform subordinates of the policy decisions made by senior management and pass up to senior management information from subordinates about problems being encountered in production.

9 *Facilitating* This involves arranging organizational conditions and methods of operation so that workers can achieve by their own efforts the organizational objectives. Thus, for example, the sales manager might allow the sales representatives a considerable amount of freedom in the planning of their visits, since he/she has found they work best this way, and provide them with contact names in companies obtained from national advertising.

Example

With what activity/activities is a production manager involved when he/she discusses with personnel the recruitment of new staff for production?

The main activities involved are liaising and deploying/staffing. You might, however, consider that the production manager is also carrying out some of the other activities listed above.

Problems

Revision questions

1 State what information is supplied by an organization chart and illustrate your answer with an example.

2 Explain what is meant by (a) flat, (b) tall, (c) matrix organizational structures.

3 Explain the difference between line and staff authority.

4 Explain how organizational structures obtained by division by function differ from those given by division by project.

5 List the main activities of a manager.

Multiple choice questions
For problems 6 to 17, select from the answer options A, B, C or D the one correct answer.

6 Decide whether each of these statements is True (T) or False (F).

Delegation means:
(i) Giving a task to someone.
(ii) Giving someone the authority to do the task.

Which option BEST describes the two statements?

A (i) T (ii) T
B (i) T (ii) F
C (i) F (ii) T
D (i) F (ii) F

7 Decide whether each of these statements is True (T) or False (F).

An organizational structure can take a number of forms. One form is a flat organizational structure. A flat organizational structure has:
(i) Many management levels.
(ii) Managers with wide spans of control.

Which option BEST describes the two statements?

A (i) T (ii) T
B (i) T (ii) F
C (i) F (ii) T
D (i) F (ii) F

8 Which one of the following is a characteristic specific to a matrix form of organizational structure?

A The business is in the form of functional departments
B A worker is accountable to more than one person
C The chain of command is not well defined
D Workers have to be non-specialists

Questions 9 to 11 relate to the following information:

The structure of organizations can be:

A Flat
B Tall
C Functional
D Matrix

Select from the above list the structure which is likely to be the most appropriate for a business which is:

9 A small business of just the owner and six staff, all located on one site

10 A company with factories at a number of widely dispersed sites around the country

11 A company which takes on large contracts for the installation of plant

12 Decide whether each of these statements is True (T) or False (F).

An organization has a functional organizational structure in which the employees are grouped according to:
(i) the product they make
(ii) the area of the country where they work

Which option BEST describes the two statements?

A (i) T (ii) T
B (i) T (ii) F
C (i) F (ii) T
D (i) F (ii) F

13 Decide whether each of these statements is True (T) or False (F).

A matrix organizational structure has employees who:
(i) are responsible to two different supervisors/managers
(ii) are grouped according to a project being worked on

Which option BEST describes the two statements?

A (i) T (ii) T
B (i) T (ii) F
C (i) F (ii) T
D (i) F (ii) F

Questions 14-17 relate to the following information:

Activities which may be undertaken by managers are:

A Planning
B Organizing
C Controlling
D Facilitating

14 Which activity would be involved when a production manager sets performance targets for his/her workers?

15 Which activity would be involved when a sales manager decides which sales staff should be responsible for different areas of the country?

16 Which activity would be involved when a production manager checks the performance of workers against that budgeted?

17 Which activity would be involved when a sales manager sets targets for the sales representatives?

Assignments

18 For either a college/school or some company, establish a chain of authority from the top to the bottom.

19 Consider two similar size businesses which use different types of production methods, e.g. one business making small batches of products and one involving mass production.
(a) Describe them in terms of their products, size and type of production methods used.

(b) Determine, in general terms, the organizational structures of the two businesses.
(c) Discuss the organizational structures in relation to the different types of production methods used.

20 Consider the public library service in your area of the country and suggest the form of the organizational structure that might be being adopted (remember there may be a County Library with perhaps branches in each main town and possibly mobile libraries for rural areas).

21 Devise possible organization charts for (a) a medium sized company making computers by assembling bought-in components, (b) a small garage business maintaining and repairing cars, (c) a large car making company with factories at a number of sites. Give reasons your organizational structures.

Case studies

22 John has been working for an organization that provides fitted kitchens. He is a plumber by trade and does the plumbing work. He decides to start up his own business installing fitted kitchens from standard units supplied by a kitchen unit manufacturer. Initially he decides just to buy-in the services of carpenters and electricians as required, doing the assembly of the units and plumbing himself. He also has to do the secretarial work, keep the accounts and do the purchasing. The orders are provided by the kitchen unit manufacturer as a result of its national publicity. After a couple of years, the workload has grown to such an extent that John takes on a carpenter, an electrician, a fitter and a secretary on a more permanent basis. After another few years the work load has increased to such an extent that the secretary needs some assistance and so John engages a clerk and a typist, putting his secretary in charge of them. The installation side has also grown and so John promotes the carpenter to production supervisor in charge of two carpenters, a plumber, an electrician and two fitters.

(a) Draw organizational charts to indicate the structure for each of the stages of John's business.
(b) How will John's activities have to change as a result of the growth of the business?
(c) A kitchen fitting business in a neighbouring town goes out of business and John decides to take over their business, so ending up with kitchen fitting businèsses in two localities. Suggest an organizational structure that John might use for this situation.

23 In 1905 a small drapery company was set up in Birmingham. It bought-in ready made clothes and sold them to the general public. The total staff consisted of the owner and three assistants. Over the years not a great deal changed. After the second world war, in 1947, the son

of the owner decided that the small shop was not going to be able to compete with the larger stores for clothes for the general public and so decided to change the emphasis into selling clothing specially designed for work, i.e. overalls, etc. This was successful and the company was able to expand by opening shops in London and Hull. In the late 1950s the business began to have difficulties obtaining the specialist clothing to the required quality and set up a factory in Walsall to make such clothing. This was very successful and the business began to sell clothing to more than just its own shops. In fact, by 1970 the factory had to be expanded to cope with demand and the number of shops had been increased to ten. The total number of workers is now about 600 in the factory and 100 in the shops.

(a) Discuss the types of organizational change that has taken place with the business over the years.
(b) Suggest the form of the organizational structure which might now operate.

24 XYZ Electronics Ltd. is a business that started in 1960 manufacturing a small range of hydraulic components for use in control engineering. The technology was changing only very slowly. The structure of the business was highly centralized. The business diversified into microprocessor controllers in 1980 by taking over another company. This business operated in a very rapidly changing technological area. This other company had a very decentralized system of authority.

(a) When the two companies merged, the managing director of XYZ Electronics Ltd imposed the same form of organizational structure and system of authority on the microprocessor controller company. Most decisions were to be made by the headquarters management at XYZ Electronics. What are the advantages and disadvantages of this?
(b) Propose an organizational structure for the combined companies.

25 ABC Electronics Ltd. is a company that was set up by Bill Smith in 1960. It then produced just small radios and consisted of Bill, a group of six assembly workers, two sales staff, a secretary and an accounts clerk. All the components were bought-in. All the staff reported directly to Bill. In 1970 the business was expanded to include the assembling of hi-fi equipment. This required a significant increase in staff numbers and for Bill to delegate authority and so he appointed a production manager and a sales manager, both reporting direct to him. However, by 1980 the volume of orders had increased to such an extent that there were 50 assembly workers all reporting to the production manager and 20 sales staff reporting to the sales manager. Bill had, since early days, kept the accounts and purchasing staff reporting direct to him. The business was, however, running into difficulties with the workers being unhappy, delays occurring with orders, production problems seeming to multiply and the managers complaining that they had insufficient time to consider

the overall strategy of the business since they had to spend all their time dealing with petty problems and issues.

(a) Draw organizational charts for the business in 1960, 1970, and 1980.
(b) Discuss possible reasons for the business running into problems.
(c) Propose a solution to the problems.

3 Business functions

3.1 Functional area activities

This chapter indicates the types of activities that are likely to occur in a number of business function areas and inter-relationships which occur with activities in other areas. The business function areas that are considered in this chapter are the key ones of finance, purchasing, marketing, sales, production, product development, quality control, stores and personnel.

The lists of activities given in section 3.1 for each of the functional areas and their inter-relationships with others are not intended to be exhaustive lists but merely to give an indication of the types of activities that might occur in a business. The inter-relationships involved in the necessary flow of information throughout a business are discussed in section 3.2.

3.1.1 Finance

The finance function involves monitoring where money is coming from and going to, when it is coming in and going out and how much. The activities include:

1 Recording all financial transactions, i.e. the details of all the money coming into the business and all the money going out, and maintaining control over the cash flow. This is termed *financial accounting* with the main records being in the form of balance sheets, profit and loss accounts and cash flow forecasts.
2 Preparing accounts to submit to customers.
3 Keeping track of all credits and debts.
4 Establishing the costs of producing the products of the business (see chapter 5).
5 Planning for expenditure within departments. This is by the production of budgets (see chapter 6).
6 Overseeing the spending within each department. This is via budgetary control (see chapter 6).
7 Planning for expenditure on capital items such as plant, machinery, equipment, buildings, etc.
8 Allowing for the depreciation of capital items, i.e. the decrease in value with time of assets of the business such as plant, machinery, equipment, furniture, etc.
9 Administering the payment of wages and salaries.
10 Ensuring that accurate records are kept of the businesses accounts. The term *auditing* is used for the checking of the accounts to ensure accuracy.
11 Preparing financial data and statistics for management planning and decision making, e.g. how the business has been performing over the last year.

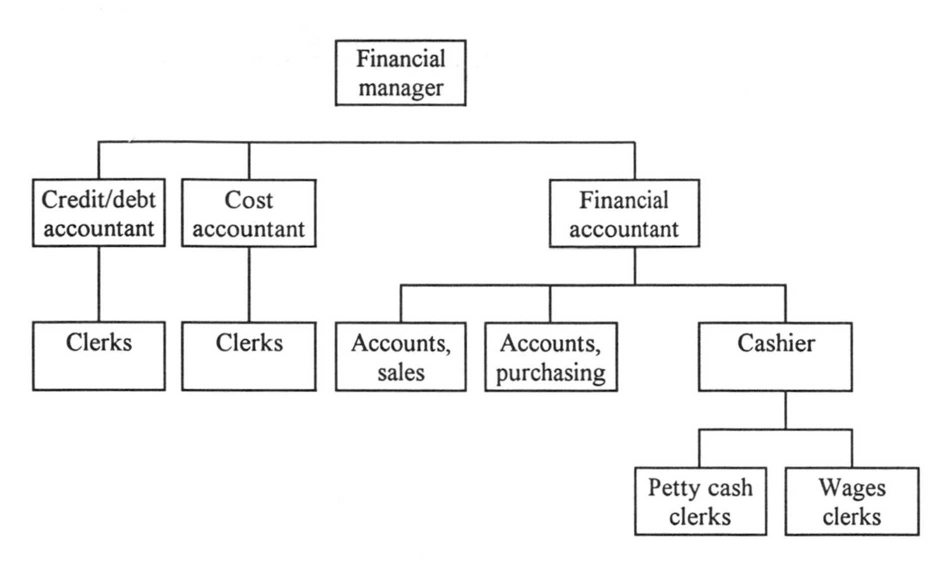

Figure 3.1 Finance department structure

The organizational structure of a finance department in a medium size manufacturing business might thus look like that shown in figure 3.1.

The inter-relationships with other functional areas are likely to include such activities as:

1 *All other departments* Payment of wages and salaries. Preparation of departmental budgets.
2 *Purchasing* Payment for the goods and services ordered.
3 *Sales* Preparation of invoices for goods and services supplied to the customers. Receiving payments.

3.1.2 Purchasing

The purchasing function is concerned with obtaining the raw materials, components and other supplies that the business needs. It typically involves the following activities:

1 Investigating sources of supply, prices and delivery times for the raw materials, components, etc. used in the business. This can involve such activities as meeting sales representatives, attending exhibitions, monitoring market trends, etc.
2 Obtaining quotations from suppliers for the materials, components, etc.
3 Negotiating terms for the supply of the raw materials, components, etc.
4 Matching delivery times of the raw materials and components, etc. with production schedules.
5 Placing orders for the purchase of raw materials, components, etc. required for production, maintenance of the plant and buildings, and administration.

6 Keeping track of orders and chasing up late deliveries of stock.
7 Monitoring performance of existing suppliers.

The organizational structure of a purchasing department in a manufacturing business might be of the form shown in figure 3.2.

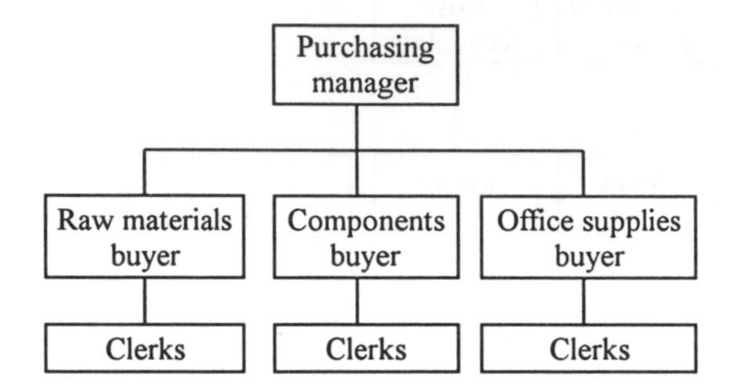

Figure 3.2 Purchasing department structure

The inter-relationships with other functional areas are likely to include such activities as:

1 *Production* Provision of the raw materials and components to the required specifications, at the right times and in the right quantities.
2 *Stores* Informing them of incoming stock and delivery dates and receiving, in return, details of stock levels.
3 *Quality control* Liaison over the quality monitoring of incoming stock.
4 *Finance* Passing invoices to finance for payment of suppliers when the account is received from the supplier.

3.1.3 Marketing

The marketing function involves establishing what the customer wants and promoting the products of the business. It typically involves the following activities:

1 Carrying out market research to establish what the customer and prospective customer wants and the demand for the products.
2 Checking the market for opportunities for increasing sales of existing products or increasing the range of products.
3 Monitoring trends to discern what the market might need in future years.
4 Watching the activities of competitors.
5 Promoting the products through advertising, mail shots, displays, etc.

The organizational structure of a marketing department in a manufacturing business might be of the form shown in figure 3.3. In some businesses, marketing and sales might be combined in a single department.

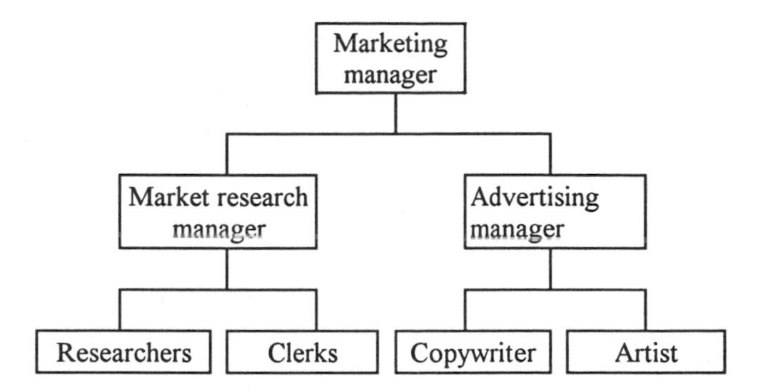

Figure 3.3 Marketing department structure

The inter-relationships with other functional areas are likely to include such activities as:

1 *Production* Supplying information to production regarding the anticipated workload and products required.
2 *Finance* Providing market information regarding numbers likely to be ordered to assist in determining pricing policy for products.
3 *Sales* Identification of the markets which can be exploited by the sales force. Receiving feedback from the sales representatives on customer reactions.
4 *Product development* Identifying to product development markets for new or redesigned products.

3.1.4 Sales

The sales function involves sales staff dealing directly with customers to sell the products of the business. This might be via the use of sales representatives with local area responsibilities. It involves such activities as:

1 Developing and maintaining contacts with customers.
2 Obtaining orders and negotiating contracts.
3 Keeping the customer informed of new products, changes to existing products, prices, etc. Providing brochures, promotional material, price lists to customers.
4 Providing technical advice to the customer.
5 Providing, by the sales office, a point of contact for customers.
6 Maintaining sales records.
7 Maintaining customer records.
8 Ensuring effective handling of customer complaints.
9 Ensuring after-sales service

The sales department is often divided into groups of sales staff, each dealing with a particular area of the country or perhaps home and export sales. A sales department might have the structure shown in figure 3.4.

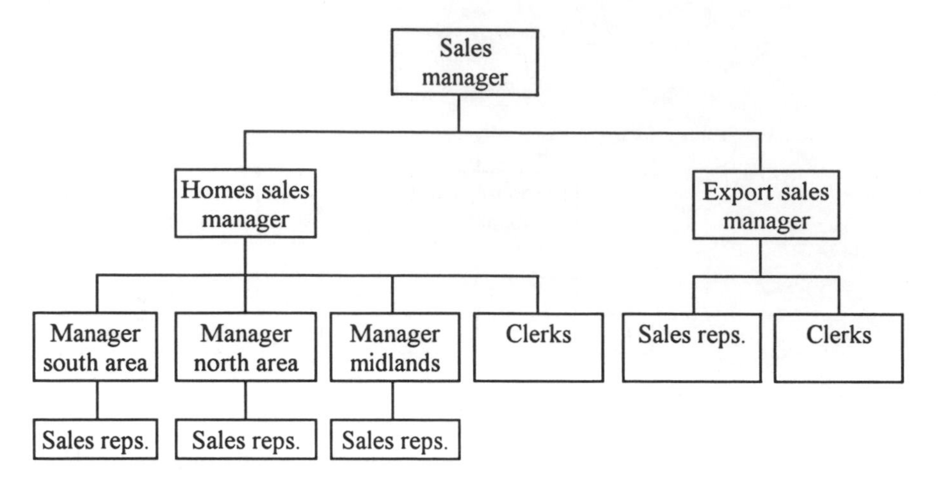

Figure 3.4 Sales department structure

The inter-relationships with other functional areas are likely to include such activities as:

1 *Production* Liaison with the production department concerning delivery dates, viable production quantities, quality, product technical information, etc.
2 *Marketing* Providing an input of information from customers to the market research.

3.1.5 Production

The production function is concerned with the making of the products, or providing the services, that are to be sold by the business. These need to be of the right quality, in the right amounts at the right times. Production planning and control and liaison with quality control are thus an essential feature. It involves such activities as:

1 Establishing production schedules to enable agreed targets to be met.
2 Translating orders into instructions for work to be carried out.
3 Scheduling incoming orders into the existing work load.
4 Checking that work is proceeding according to plan.
5 Carrying out production schedules to produce the products.
6 Controlling the costs of workers, machines and materials in order to achieve the desired output at minimum cost.
7 Keeping records of labour, machines, materials used so that jobs can be costed.
8 Planning tool changes and tool settings.
9 Planning for the installation of new equipment.
10 Installing new equipment.
11 Maintaining the equipment used in production.

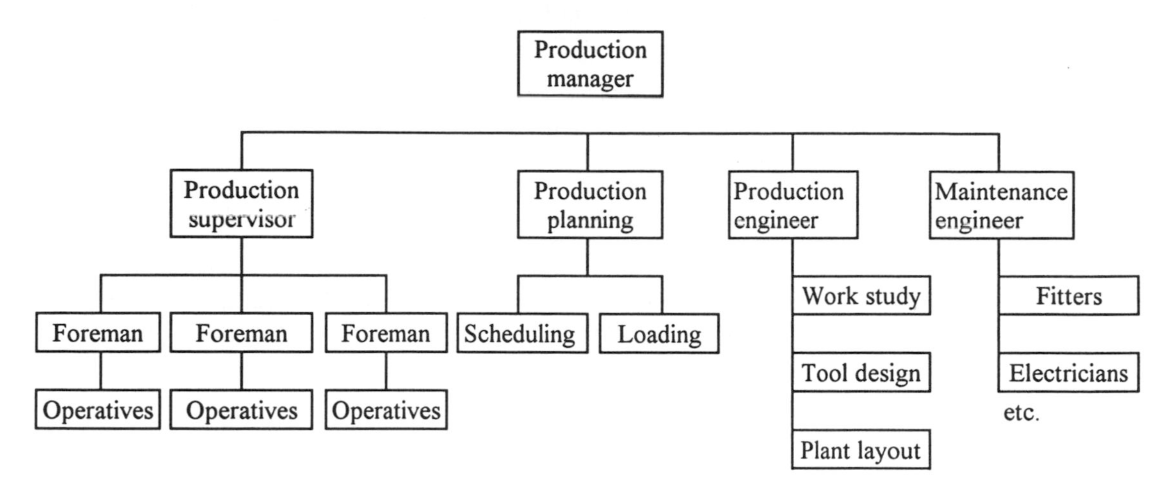

Figure 3.5 Production department structure

The organizational structure of a production department in a manufacturing business might be of the form shown in figure 3.5. Also part of a production department might be the stores and those aspects of quality control concerned with the product.

The inter-relationships with other functional areas are likely to include such activities as:

1 *Stores* Drawing materials and components from stores. Supplying finished products to stores.
2 *Purchasing* Ensuring that the right materials and components are ordered in the required quantities and available at the required times.
3 *Quality control* Ensuring that the production output is of the right quality.
4 *Finance* Providing operation times and work details for wage and cost calculations. Providing equipment replacement, new equipment, new staffing, etc. proposals for financing.
5 *Marketing* Establishment of production scheduling plans to meet marketing forecasts.
6 *Sales* Supplying technical information, delivery data, etc. regarding products.
7 *Personnel* Liaison with personnel regarding staff training, health and safety, labour recruitment and selection, etc.
8 *Product planning* Liaison with product planning in their preparation of plans for new or revised products.

3.1.6 Product development

The product development is often included under the general heading of research, development and design. It is concerned with the development of new products and the revision of existing products. As such it has to have a close relationship with both marketing and production. It involves such activities as:

1 Keeping up with latest developments in materials and processes.
2 Research into new materials and processes relevant to the products of the business.
3 Development of new and revised products.
4 Development of new and revised production methods.
5 Design of new and revised products.

The organizational structure of a product development department in a manufacturing business might be of the form shown in figure 3.6.

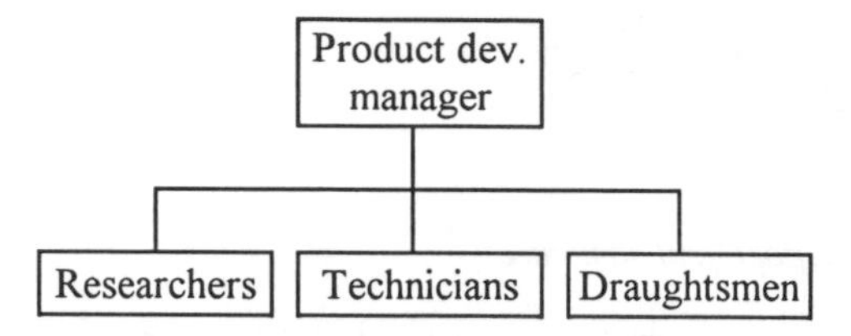

Figure 3.6 Product development department structure

The inter-relationships with other functional areas are likely to include such activities as:

1 *Production* Liaison over the development of new and redesigned products and processes.
2 *Marketing* Liaison over new and redesigned products to be marketed and the requirements of existing and potential customers.
3 *Sales* Designing new and improved products to enable sales to beat the competition.
4 *Finance* Providing details of materials, processes, and the implications in general of new or revised designs on costs.

3.1.7 Quality control

The term quality when applied to a product or service produced by a business implies that it is fit for the purpose for which it is sold to the customer. Thus if a customer buys, say, a new car he or she expects to be able to drive the car without the bodywork deteriorating or the engine malfunctioning for at least a year. The quality of the products and services offered by a business is of importance to both the organization and its customers. A business needs to deliver, to the customers, products and services which are to the customer's requirements and so fit for the purpose for which they were bought. A business which fails consistently to do this will lose business. Thus an organization needs to establish and maintain an effective system for ensuring that its products and services conform to the specified requirements of the customers. The standard laid down for a business to establish such a quality system standard throughout the entire organization is specified in the British Standard BS 5750, and in an identical way in Euro Norm EN 29000 and International Standards Organization ISO 9000.

The quality control function is concerned with ensuring that the quality of all the products and services which the business sells to customers is consistently up to the specification required by the customers. It is thus concerned with such activities as:

1 Provide inspection staff, gauges and testing equipment, to test incoming materials and components and the product both during manufacture and on completion of manufacture.
2 Testing and checking incoming materials and components to ensure that they conform to the specifications laid down in the purchasing contracts.
3 Testing and checking the products produced by the business to ensure that they conform to the specifications laid down in the sales contracts.
4 Rejecting products which do not conform to the required standards and decide which can be reworked to bring them up to standard.
5 Testing and checking the products during the production process to ensure that the production unit can be warned when the product is in danger of going outside the specified limits or prevent time being wasted on further work on products which have already failed to meet the required standards.
6 Testing and checking the handling, storage, packing and delivery of products to ensure that quality is protected.
7 Work within the quality control policy of the business.

The organizational structure of a quality control department in a manufacturing business might be of the form shown in figure 3.7. Such a department might form a section under the production manager.

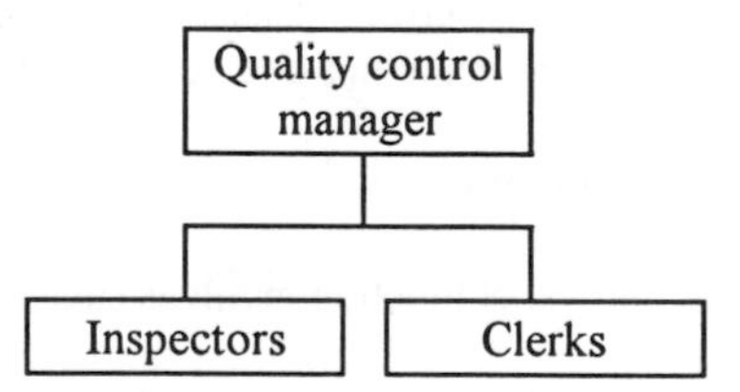

Figure 3.7 Quality control department structure

The inter-relationships with other functional areas are likely to include such activities as:

1 *Marketing* Working with marketing to establish what the customer requires, only then can quality control work to deliver the required standard.
2 *Production* Providing quality checks and early warnings when products are beginning to go outside the specified limits, so that machines can be

reset, tools replaced, etc. Indicating to production which products have to be reworked to bring them up to standard.
3 *Stores* Indicating to stores which incoming materials and components are not up to the standards specified by purchase orders.
4 *Product development* Liaison with product development to ensure that the standards specified in the design of the product are fulfilled in the finished products.
5 *Sales* Ensuring that the products sold conform to the specifications agreed by sales with the customers.

3.1.8 Stores

The stores function involves such activities as:

1 Storing and keeping safe materials and components delivered to the business by suppliers.
2 Checking the incoming materials and goods against the delivery notes.
3 Issuing materials and components against requisitions.
4 Recording stock movements into and out of store.
5 Keeping check of the stock held.
6 Storing finished products when they are to be sold from stock.

The inter-relationships with other functional areas are likely to include such activities as:

1 *Production* Providing production with the materials and components needed to production. Storing completed products.
2 *Purchasing* Being informed by the purchasing department that materials and components have been ordered and informing the purchasing department when the orders have been received and checked.
3 *Quality control* Liaison with quality control in the checking of incoming materials and components.

3.1.9 Personnel

The personnel function involves such activities as:

1 Providing written job specifications.
2 Recruiting labour.
3 Providing written employment procedures. This might include details of training schemes, incentive schemes, pensions, holidays, redundancy, disciplinary and grievance procedures, etc.
4 Devising and organizing training programmes.
5 Promoting health and safety throughout the business and implementing the Health and Safety at Work Act 1974 and other relevant safety acts and regulations.
6 Maintaining records of personnel.
7 Negotiating a wages/salary policy for the business.

8 Negotiating with the relevant trade unions concerning conditions of service.
9 Dealing with industrial relations.

The organizational structure of a personnel department in a manufacturing business might be of the form shown in figure 3.8, each officer being supported by clerical and secretarial staff.

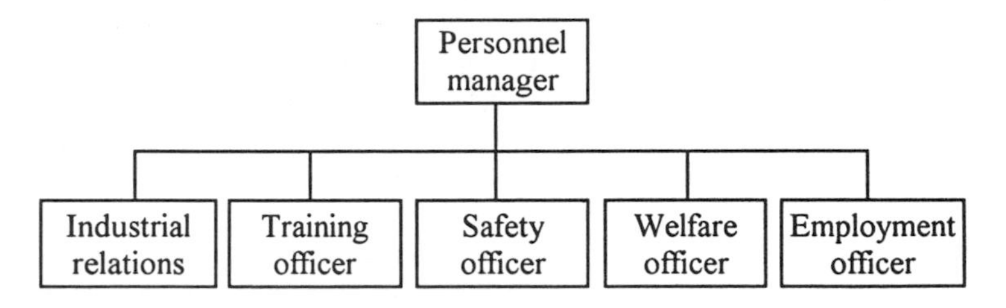

Figure 3.8 Personnel department structure

The inter-relationships with other functional areas are likely to include such activities as:

1 *All departments* Liaison with departments regarding hiring and training staff, health and safety, staff welfare, industrial relations, wages and salaries, etc.
2 *Finance* Liaison with regard to wages and salary policy and budgeting for training, staff welfare, etc.

Example

Which functions in a business might be expected to have the main responsibility for (a) inspection of incoming material and components, (b) arranging training for staff, (c) obtaining quotations from suppliers for materials?

(a) Quality control liaison with stores.
(b) Personnel liaison with the relevant department.
(c) Purchasing liaison with the production department for the specifications of what is required.

3.2 Communication interfaces

Information has to move between the various functions in a business. Thus, for example, marketing may produce forecasts of the level of demand for a product. This information has to be passed to production where it is used to plan the production facilities for this demand. This planning will affect finance, since labour and materials will have to be budgeted for. Stores will require the information since they will have to ensure that sufficient stocks of materials and components are in stock when required. Thus an input of information to one function can result in an information flow into other functions.

The following could be some of the actions involved, from sales representatives meeting a customer to delivery of the completed order of a non-standard product to the customer and the resulting payment to the business.

1 Sales representatives discuss with potential customers the product.
2 Sales representatives liaise with the sales department who liaise with production to give delivery dates.
3 Sales department liaise with finance for estimates of costs.
4 The sales function supplies the customer with a delivery date and cost.
5 The customer places an order.
6 The order is processed by the sales department.
7 Production is advised of the order.
8 Finance is advised of the order.
9 Production considers how plant and labour are being utilized and schedules the order into the existing work load.
10 Production advises purchasing of the materials and components that have to be bought.
11 Purchasing obtains quotations and then orders the materials and components.
12 Stores is advised of the purchase order.
13 Stores receives the materials and components ordered and checks them, assisted by quality control.
14 Production requisitions materials and components from the stores.
15 Production makes the product.
16 Quality control checks the product.
17 Production delivers the product to stores/dispatch and it is then dispatched to the customer.
18 Dispatch advises finance that the order has been dispatched.
19 Finance bills the customer.
20 Finance receives payment for the order.

Consider another example. The marketing section, from a market survey of potential customers, discern a need for a new product. This information has then to be passed to the product development function to produce a design for the new product. This then starts a whole sequence of information flows and actions before the product is finally launched. Integral with this sequence will be a consideration by senior management of the implications of making the new production and whether it conforms with their overall policy and can be justified on financial grounds. The actions involved, though not necessarily precisely in the sequence indicated, might be:

1 Market survey by marketing function discloses the need for the new product and the type of specification required.
2 Finance consider the potential costs and benefits to the business of making and selling such a product.
3 Senior management consider the implications of making such a product.
4 A prototype is authorized.

5 Product development design the product.
6 Product development produce a prototype of the initial design.
7 Marketing market test the prototype.
8 Product development revise the product in the light of the tests.
9 Product development liaise with production to produce drawings and specifications for the new product.
10 Finance cost the product and budget for production.
11 Purchasing obtain quotations for materials and components that are to be bought.
12 Production plan the production methods and schedule.
13 Production design and manufacture tools for the new product.
14 Purchasing order the materials and components required for production.
15 Personnel in conjunction with production train production workers, recruiting extra if necessary.
16 Stores receive the materials and components ordered for the new product.
17 Production requisition materials and components from stores.
18 Production carry out a trial production run to iron out any teething problems.
19 Production manufacture the new product.
20 Stores receive the new product for store.
21 Marketing design promotional materials for the new product.
22 Marketing promote the new product.
23 Sales representatives advise customers of the new product.
24 Sales obtain orders.

Example

The sales department in a business have received an order for a large quantity of a standard product manufactured by the business. What are likely to be their initial contacts with other departments in the business?

Since the product is a standard one they would most likely contact stores to find out the situation regarding stocks of that product and whether the order could be met from stock. If this is not possible then production would have to be involved in order to consider how the order could be met. Finance would need to be informed of the receipt of the order and also might be involved in considering the implications of production taking on the extra production. These are just a few of the possible contacts that might occur, you can no doubt think of more.

3.2.1 Communication chains

Consider a group of people, call them A, B, C, D, and E, and possible methods of communication within the group. We might have the situation that A is the group leader and all communications effectively flow to or from A. The communications can then be considered to follow the paths shown in figure 3.9(a). Alternatively we could have a group situation where

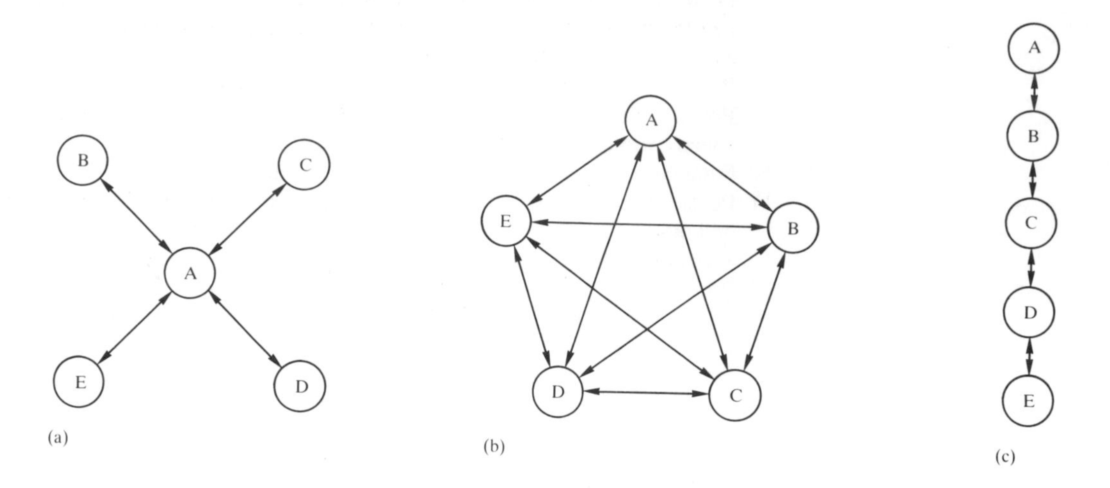

Figure 3.9 (a) Person A dominates communication, (b) no dominant person in the group, (c) a linear chain of communication

everbody communicates equally with everybody else and there are no dominant communication paths. The situation might then be described by figure 3.9(b). Another possibility is that A communicates with B who then communicates with C who then communicates with D who then communicates with E, figure 3.9(c) illustrating the communication paths in this case.

Different forms of communication chains have different characteristics. Thus where all the communication flows through one person, i.e. the arrangement described in figure 3.9(a), has the advantage of being a quick and accurate way of communicating and allowing A to control the situation. This might describe communications between a supervisor and four workers or a managing director and four senior managers. With the system described by figure 3.9(b) there is a completely decentralized structure with everybody contributing equally. This implies no leader. It is an effective method of communication where a problem has to be solved and all minds bear on the problem. The linear communication chain described in figure 3.9(c) is a slow method of communicating information and can also be a poor method. It is slow since the information has to pass through many people. As some misinterpretation might occur at any stage, it can also be a poor method with the communication arriving at the end of the chain becoming distorted. This last type of communication chain is common in organizations where a senior manager communicates to a middle manager who communicates to a supervisor who communicates to a worker.

Within any organization there are likely to be a number of different types of communication chains. The pattern will obviously depend on the degree of centralization adopted by the organization, the communication paths often, though not always, following the lines of authority. The term *vertical communication* is used to describe the communication that occurs in an organization from top decision makers down to the employees at a lower level who have to implement the decisions and also, in reverse, the communications from the employees up to the decision makers. The term *horizontal communication* is used to describe the communication that occurs between employees who operate at similar levels in the organization. Vertical communications in an organization tend to occur in a planned manner which is linked to the structure of the organization. Horizontal communication might not always be planned.

To illustrate the ideas of vertical and horizontal communications in a business, figure 3.10 shows some of the communications that can occur.

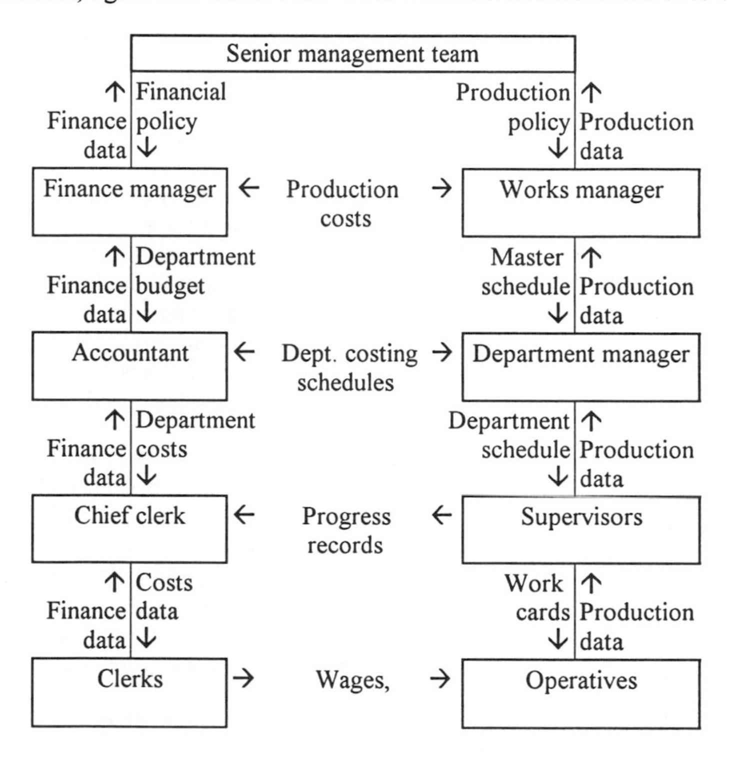

Figure 3.10 Examples of vertical and horizontal communications

Example

What type of communication chains might be present in the following situations in a college/school:

(a) Communications between the lecturer/teacher and the students.

(b) Communications between students when a group of them are undertaking a practical exercise/project.

(c) Communications between the college principal/school head and the students in some class.

(a) This is likely to be of the form shown in figure 3.9(a).
(b) This is likely to be of the form shown in figure 3.9(b), if we assume that there is no one dominant person in the group.
(c) This is likely to be of the form shown in figure 3.9(c) with communications following a line down from the principal/head.
Revision questions

Problems

1 List the typical functions of (a) sales representatives, (b) storekeepers, (c) engineers in production planning ,(d) safety officers.

2 Explain what is meant by quality control.

3 Outline a possible structure for a production department.

4 List the engineering activities that might typically be involved in a production department.

5 Give one example in each of the following of the type of information that might pass between:
(a) marketing and product development,
(b) quality control and production,
(c) finance and production,
(d) purchasing and stores,
(e) production and stores,
(f) product development and production.

6 Explain what are meant by vertical communication and horizontal communication in an organization and give an example of each.

Multiple choice questions
For problems 7 to 17, select from the answer options A, B, C or D the one correct answer.

Questions 7-9 relate to the following information:

Four examples of the activities carried out by staff in a business are:

A Negotiating a loan
B An advertising campaign for a new product
C Checking the quality of incoming materials
D Scheduling incoming orders against the existing work load

Select the most relevant business function which would be directly concerned with the above activities:

7 Marketing
8 Production
9 Finance

10 Materials and components needed for production must be provided at the right time and in the right quantities to complete orders to meet delivery dates.

Decide whether each of these statements is True (T) or False (F).

It is the responsibility of the purchasing function to:
(i) Place the orders with suppliers.
(ii) Translate orders into requirement for materials and components.

Which option BEST describes the two statements?

A (i) T (ii) T
B (i) T (ii) F
C (i) F (ii) T
D (i) F (ii) F

11 Decide whether each of these statements is True (T) or False (F).

It is the responsibility of the sales function to:
(i) Develop new products.
(ii) Develop new outlets for products.

Which option BEST describes the two statements?

A (i) T (ii) T
B (i) T (ii) F
C (i) F (ii) T
D (i) F (ii) F

Questions 12-15 relate to the following information:

The following are some of the business functions that might be present in a business:

A Production
B Marketing
C Finance
D Personnel

12 Which business function is responsible with familiarizing new staff with the conditions of service?

13 Which business function is responsible with establishing what the customer wants?

14 Which business function is responsible for costing products?

15 Which business function is responsible for developing training for employees?

16 Market research has disclosed the need for revisions to one of the products produced by a company.

Decide whether each of these statements is True (T) or False (F).

The marketing department will need to discuss the outcomes of the market research with:
(i) Product development
(ii) Finance

Which option BEST describes the two statements?

A (i) T (ii) T
B (i) T (ii) F
C (i) F (ii) T
D (i) F (ii) F

17 The sales department has received an order for items from the list of standard products produced by the company.

Decide whether each of these statements is True (T) or False (F).

The sales department will need to inform:
(i) Stores
(ii) Finance

Which option BEST describes the two statements?

A (i) T (ii) T
B (i) T (ii) F
C (i) F (ii) T
D (i) F (ii) F

Assignments

18 For a college/school determine the activities required from the maintenance function and outline the structure of that section.

19 List the activities that could be involved when you place an order with a car showroom for a new car which is to be delivered from stock held by the car manufacturer.

Case studies

20 The XYZ Company makes products X, Y and Z. The production department has three production lines which produce the three products

at a steady rate. The company has run into financial problems which have been diagnosed as arising from the company carrying too large a stock of finished products. In the case of product X the stock is way beyond what might forsccably be sold in the near future. Products Y and Z stocks are high, but not so excessively. The managing director feels that this arises from a lack of proper communication between departments in the company. The following are some details concerning the structure of the company:

Departments in the company:

Accounts: This deals with the company finances, the payroll, invoices to customers, customer credit.
Sales: This is the department having direct contact with customers, generally through sales representatives.
Purchasing: This department buys the raw materials used by production.
Manufacturing: This department makes and assembles the three products.
Production control: This department has the job of determining what manufacturing should make and has information about current stock levels, orders and production capacity.
Stores: This department store the raw materials and the finished products.
Dispatch: This department packs the products and sends them to customers.

(a) List the information flow pattern that is likely to occur when orders are received.
(b) Identify problems in the information flow pattern that might be responsible for the overstocking.
(c) Suggest reasons for the problems and hence solutions which might be considered.

21 The ABC Engineering company is a medium size company which makes plastic tubing. The company is divided into three departments; production, finance and administration. The administration department provides as a general service for the company a secretarial pool, filing and reprographics, the mail room service, switchboard operator and reception, security guards, and covers the personnel function. The department operates under an administration manager, with the secretarial pool supervisor, the clerical supervisor, the senior mail room clerk, the senior security officer and the personnel officer reporting directly to him/her. The two receptionists/switchboard operators report to the senior clerical officer.

(a) Sketch the organization chart of the administration department.
(b) List the activities that might be expected of the administration department.

(c) Determine the communication path that is likely when an important visitor complains to the managing director about the sloppy treatment he receives from reception.
(d) The company makes plastic tubing of a range of standard diameters, colours and lengths. To do this it purchases raw plastics materials in bulk and stores them pending withdrawals from store for production. The company has a number of extruders and the material is then transported from the store to whichever of them needs supplies. The material is then fed into the machines which extrudes continuous tube. This is cut into suitable lengths and collected in batches. Inspectors take samples from each batch and test them for quality. The batches, if passed by inspection, are then transported to a finished goods store, where they await orders. List a possible sequence of events for when a large order is received and cannot be met entirely from tubing in stock.

22 A group of students decide that they could make money by selling summary lecture notes for key subjects in the course they are taking. The three main motivators decide that they will be in charge, one of them assuming responsibility for marketing, and sales, another having the responsibility for the writing and production of the notes, while the third has responsibility for finance.

(a) Work out the possible activities for which each of the three are responsible.
(b) Work out a possible sequence of events that could occur from orders being sought and gained to the delivery of the order to the customer.

4 Financial factors

4.1 Financial factors

This chapter is a discussion of financial factors in relation to business decisions. Such decisions are affected by consideration of factors such as operational costs, capital costs, investment costs, and the relation of costs to the income a business will receive from sales.

As an example of such factors, consider a manufacturing business. The types of questions that might occur are:

How many items will have to be made in order to make a profit?

In connection with the installation of equipment there can be a number of questions, e.g. should the business install sand casting equipment or die casting equipment? How do the capital costs of installing sand casting compare with those of die casting? How do the operational costs of sand casting compare with those of die casting?

Would it be cheaper to buy-in a component rather than make it in-company?

How large a stock of raw materials/work-in-progress/finished goods should be held?

A way of considering such questions is discussed in this chapter.

4.1.1 Fixed and variable costs

The term *capital cost* is used for a cost which has to be paid for up-front for equipment or premises, etc. before any income is received. It is a *fixed cost* and is independent of whether the equipment, etc. is used or not. The decision to incur a capital costs is an *investment decision* which involves an outlay of cash now in return for benefits in future years. The term *operational cost* is used for the costs involved in producing items. It is the cost of raw materials, components that are used up in making the product, maintenance costs, etc. While there may be some element which is a fixed cost, it is generally a *variable cost* and will often increase in proportion to the number of items produced.

Thus we can consider costs to be divided into fixed costs and variable costs. What constitutes such costs is discussed in more detail in chapter 5.

1 *Fixed costs* stay unchanged, not changing with volume over some range in the volume of production or volume sold.
2 *Variable costs* are those costs which change with the volume of production or volume sold.

Suppose a business has, for a particular product, fixed costs of £6000 and variable costs of £2 per item produced. The cost of producing 100 items will thus be

cost of 100 items = fixed cost + 100 × variable cost per item

Thus

cost of 100 items = 6000 + 100 × 2 = £6200

Now consider what the cost would be of 1000 items of this product.

Cost of 1000 items = fixed cost + 1000 × variable cost per item

= 6000 + 1000 × 2 = £8000

Example

The costs incurred by a company in producing a product are a fixed cost of £10 000 and a cost of £4 per item produced. What will be the cost of producing (a) 100, (b)1000 items?

(a) As above, we thus have

cost of 100 items = fixed cost + 100 × variable cost per item

= 10 000 + 100 × 4 = £10 400

(b) In a similar manner we have

cost of 1000 items = fixed cost + 1000 × variable cost per item

= 10 000 + 1000 × 4 = £14 000

4.2 Operational costs

Consider the problem of deciding which process should be used to manufacture some item. Suppose the choice is between, say, sand casting or die casting. If we restrict our consideration to just the *operational costs*, then the argument might proceed as follows. Sand casting involves the making of a mould using a mixture of sand with clay. With sand casting a new mould has to be made for each item manufactured. Thus the cost of the mould increases in proportion to the number of items made. The cost of making such a mould is, however, comparatively low. Die casting involves the making of a metal mould. With die casting the same mould can be used for a large number of items. The cost of the mould is thus fixed and does not increase as the number of items made increases. The cost of making such a mould is, however, much higher than the cost of making a sand mould. Thus considering just the mould costs, the situation might be as described by figure 4.1.

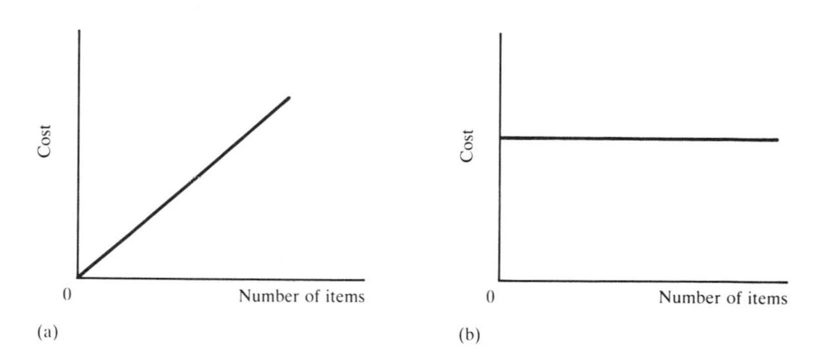

Figure 4.1 Mould costs for (a) sand casting, (b) die casting

To the mould costs we must add raw material costs, labour costs, etc. These costs will increase with the number of items made. Thus the total costs of using sand casting and die casting might be as shown in figure 4.2. Thus if there was an order for say a small number of items then sand casting would have the smaller operational costs. However, if the order was for a large number of items then the initial cost involved in making the die would be spread over the larger number of items and the cost per item with die casting could be smaller than that for sand casting.

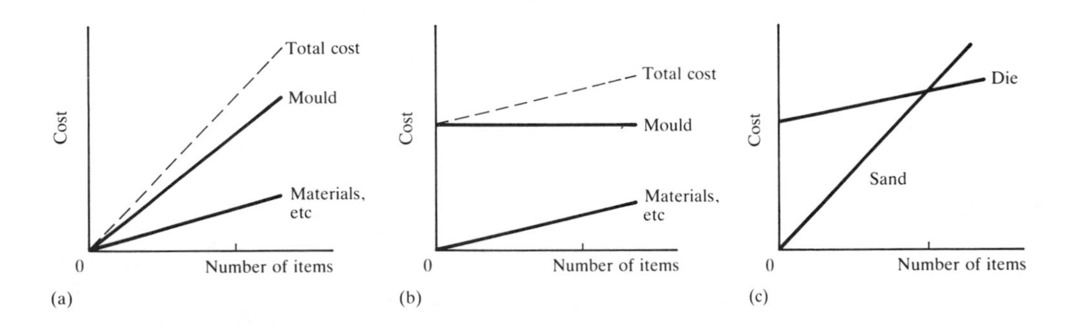

Figure 4.2 Operational costs for (a) sand casting, (b) die casting, (c) sand and die casting compared

4.3 Capital costs

When, for example, a new machine is purchased then capital costs are being incurred. After one year the resale value of the machine will be less than the price paid for it. The difference is the cost which has to be defrayed against the products made by the machine in that year. Thus if, for example, a new machine cost £50 000 and after one year its resale value was £35 000, then £15 000 is the capital cost element that must be defrayed against the products made by the machine in that year. Thus if, say, 10 000 items were made then there would be a capital cost element of £1.50 made against each item.

Consider the problem of determining which equipment should be installed in a production department. Suppose the equipment is to make products from a thermoplastic, and the products will be hollow and of fairly simple form. The choice might then be between injection moulding and rotational moulding (other methods are possible, but for the purpose of this example, only these two alternatives are considered). With injection moulding the polymer is melted and forced into a mould. With rotational moulding the powdered polymer is melted inside a heated mould which is rotated. Injection moulding can produce many more items per hour than rotational moulding, e.g. it might be 50 per hour for injection moulding compared with two per hour for rotational moulding. Here we will consider the capital costs involved. Thus we might have the cost of buying and installing the equipment for injection moulding as £90 000 and the cost for rotational moulding as £10 000.

Capital costs for installations are usually defrayed over an expected lifetime of the installation. However, in any one year there will be a depreciation of the asset and this is the capital cost that is defrayed against the output of the product in that year. Thus with the capital expenditure needed to purchase a machine and install it for £90 000 then in one year a depreciation factor of 10% might be used and thus, in the first year, £9000 is defrayed as a capital cost against the quantity of product produced in that year. Thus if, say, the injection moulding machine was used to produce 100 000 items in that year then against each item the capital cost element would be £0.09. For the rotational moulding machine with an initial cost of £10 000 and a depreciation of 10% then £1000 is defrayed as a capital cost against the quantity of product produced in the first year. If the rotational moulding machine was used to produce 5000 items in a year, then against each item the capital cost element would be £0.50. Table 4.1 shows how the value of the machines might vary with time if a depreciation of 10% per year is assumed.

Table 4.1 Example of depreciation of 10% per year

		Value of asset after:		
	Initially £	*1 year* £	*2 years* £	*3 years* £
Injection moulding machine	90 000	81 000	72 900	65 610
Rotational moulding machine	10 000	9 000	8 100	7 290

There are other factors which would need to be considered before a decision can be made regarding which machine to install. For example, the die cost for the injection moulding is, say, £2000 and that for rotational moulding £500. These have to be specially made for each product. Thus if both machines are to be used to produce, say, 2000 items then the cost per item of the dies is £1 for injection moulding and £0.25 for rotational moulding.

Example

A business is considering buying a machine which costs £40 000 to buy new. The expected end-of-year resale value for the machine after 1 year is £28 000, after 2 years is £18 000, after 3 years is £10 000 and after 4 years is £4000. What will be the capital cost element that should be defrayed against the quantity of product produced in each of the years?

In the first year the capital value of the machine declines by £12 000 and so this is the amount which has to be defrayed against the product costs for that year. In the second year the capital value declines by £10 000, in the third year by £8000 and in the fourth year by £6000. These are thus the amounts which have to be defrayed against the product costs in those years.

4.4 Costs and revenue

The costs of production can be considered to be made up of two elements: fixed and variable costs. *Fixed costs* are those costs which do not increase as the number of items produced increases and include such costs as an element of the capital and factory building costs. *Variable costs* are those costs which depend on the number of items made. These include material costs, labour costs, energy costs, etc. (see chapter 5 for more discussion of fixed and variable costs). Suppose the variable costs increase in proportion to the number of items made. The total cost per item sold is given by

total cost = variable cost + fixed cost

The variable cost is the product of the variable cost per item and the number of items sold. Thus

total cost = (number of items sold × variable cost per item) + fixed cost

The total cost might thus vary with the number of items sold in the way shown in figure 4.3.

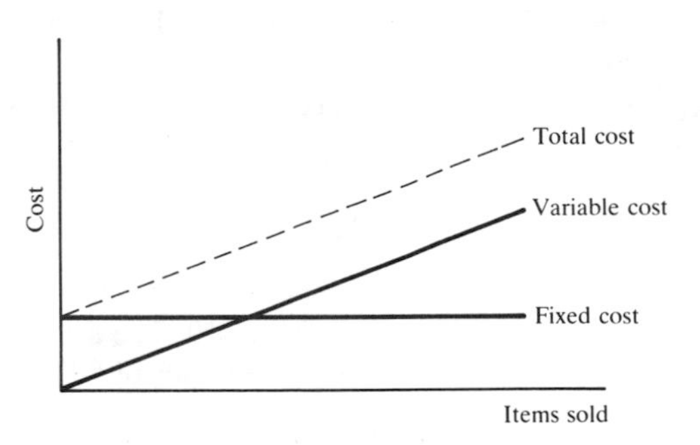

Figure 4.3 Costs relationship with sales

Costs are incurred by a business in order that sales can be made and sales revenue obtained. Thus a business in making decisions about, say, installing new machinery or undertaking an order will be concerned with the relationship between the costs and the sales revenue. The sales revenue will depend on the number of items sold and the price charged per item sold. Thus we might have

sales revenue = number of items sold × item selling price

and so a sales revenue relationship with sales of the form shown in figure 4.4. Such a relationship assumes that the selling price per item is fixed and no discounts are offered for quantity.

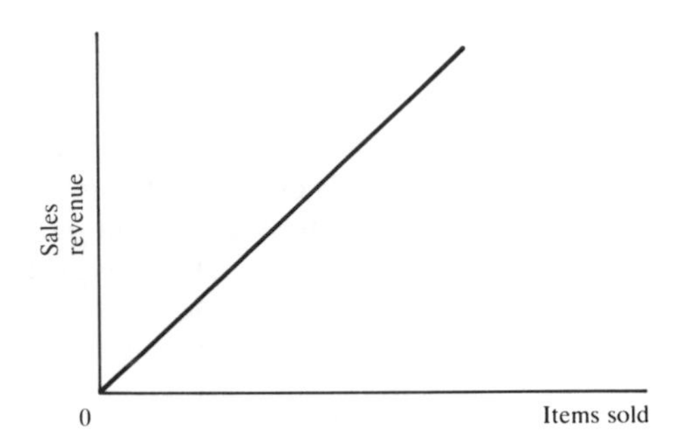

Figure 4.4 Sales revenue and sales

The profit at any particular volume of sales is the difference between the sales revenue and the total cost, i.e.

profit = sales revenue – total cost

If the total cost is greater than the sales revenue then the profit is a negative value and so is a loss. Thus the profit is

profit = number of items sold × item selling price
– (number of items sold × variable cost per item) – fixed cost

Example

A product has a fixed cost per year of £30 000 and a variable cost per item of £10. The selling price per item is £20. What will be the profit if there are sales of 5000 per year?

The total costs per year for a production of 5000 items is

total costs = 5000 × 10 + 30 000 = £80 000

The total sales revenue per year is

$$\text{sales revenue} = 5000 \times 20 = £100\,000$$

Thus the profit is

$$\text{profit} = \text{sales revenue} - \text{total costs}$$

$$= 100\,000 - 80\,000 = £20\,000$$

Example

A product has a fixed cost per year of £30 000 and a variable cost per item of £10. The selling price per item is £20. How many items need to be sold per year for a profit of £40 000 to be made?

Using the equation

profit = number of items sold × item selling price
− (number of items sold × variable cost per item)
− fixed cost

then

$$40\,000 = n \times 20 - n \times 10 - 30\,000$$

Hence

$$10n = 70\,000$$

and so the number n of items to be sold to make such a profit is 7000.

Example

A product has a fixed cost per year of £30 000 and a variable cost per item of £10. What should the selling price be for the product to show a profit of £20 000 on sales of 4000 items?

Using the equation

profit = number of items sold × item selling price
− (number of items sold × variable cost per item)
− fixed cost

$$20\,000 = 4000 \times \text{item selling price} - 4000 \times 10 - 30\,000$$

Hence

$$\text{item selling price} = \frac{20\,000 + 40\,000 + 30\,000}{4000} = £22.50$$

4.4.1 Break-even analysis

The two relationships, cost/items sold and sales revenue/items sold, can be combined on a graph to give information on the effect of producing those items on profits. This method is called *break-even analysis*. Figure 4.5 shows the cost/items sold graph from figure 4.3 and the sales revenue/items sold graph from figure 4.4 on the same graph axes. The point where the two lines cross is the *break-even point*. Below that number of items sold the costs are greater than the sales revenue. Above that point the sales revenue is greater than the costs.

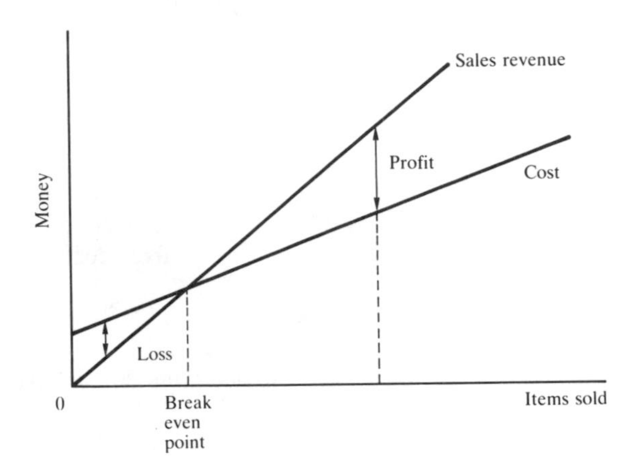

Figure 4.5 Break-even chart

The profit at any particular volume of sales is the difference between the sales revenue and the total cost, i.e.

profit = sales revenue − total cost

Thus on the break-even chart the profit at a particular volume of sales is represented by the line shown in figure 4.5. The equation for the cost line in figure 4.5 is given by the equation developed in the previous section, i.e.

total cost = (number of items sold × variable cost per item) + fixed cost

The equation for the sales revenue line is given by the equation developed for sales in the previous section, namely

sales revenue = number of items sold × item selling price

The break-even point is where these two lines cross, i.e. when the total cost equals the sales revenue, i.e. there is zero profit. Thus we must then have

(number of items sold × variable cost per item) + fixed cost
= number of items sold × item selling price

This equation can be rearranged to give

fixed cost = number of items sold × item selling price
− (number of items sold × variable cost per item)

and so the number of items sold at the break-even point is given by

$$\text{number of items sold} = \frac{\text{fixed cost}}{\text{item selling price} - \text{variable cost per item}}$$

Example

The costs and revenue for a particular product are:

Fixed cost £2000
Variable cost £4 per item
Selling price £9 per item

Determine the cost of producing, the revenue obtained, the profit/loss when (a) 100, (b) 200, (c) 300, (d) 400, (e) 500, (f) 600 items are produced. Hence plot the break-even chart. What is the break-even volume of sales?

The cost of a number of items is given by

cost = fixed cost + number of items × cost per item

and the revenue by

revenue = number of items × selling price per item

The profit or loss is the difference between the revenue and the cost. Thus, using the above data:

Number of items	*Fixed cost £*	*Variable cost £*	*Total cost £*	*Revenue £*	*Profit £*	*Loss £*
100	2000	400	2400	900		1500
200	2000	800	2800	1800		1000
300	2000	1200	3200	2700		500
400	2000	1600	3600	3600		Break even
500	2000	2000	4000	4500	500	
600	2000	2400	4400	5400	1000	

Figure 4.6 shows the above data plotted on a break-even chart. The break-even point, i.e. where the sales revenue equals the costs, is a volume of 400 items.

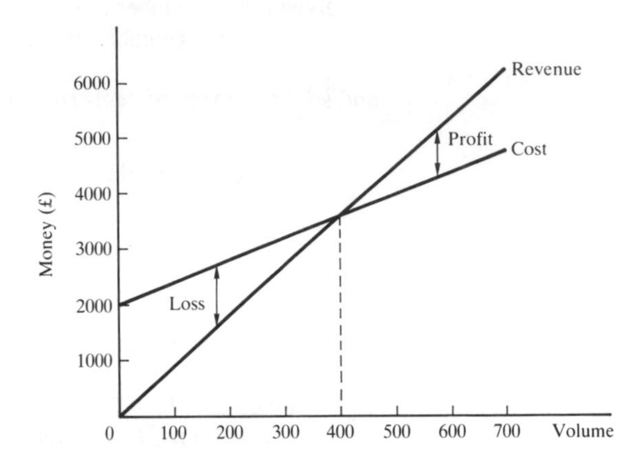

Figure 4.6 Example

Example

A product has a fixed cost per year of £30 000 and a variable cost per item of £10. The selling price per item is £20. How many items need to be sold per year for the business to break even?

$$\text{number of items sold} = \frac{\text{fixed cost}}{\text{item selling price} - \text{variable cost per item}}$$

$$= \frac{30\,000}{20 - 10} = 3000$$

Example

A product has a fixed cost per year of £30 000 and a variable cost per item of £10. The selling price per item is £20. What will be the effect on the break-even volume if the selling price is reduced by 10%?

Before the selling price is reduced the break-even volume is as obtained in the above example, namely 3000. A reduction of the selling price by 10% means a reduction by 10% of £20, namely £2. Thus the new selling price would be £18. The break-even volume is then

$$\text{number of items sold} = \frac{\text{fixed cost}}{\text{item selling price} - \text{variable cost per item}}$$

and so

$$\text{number of items sold} = \frac{30\,000}{18 - 10} = 3750$$

Thus the break-even volume is increased by 750. This is a percentage increase of $(750/3000) \times 100 = 25\%$.

4.4.2 Break-even analysis applied to a production problem

As an example of the application of break-even analysis, consider a production manager faced with the decision as to what type of production layout might be suitable for a particular product. One of the possibilities might be that the machines for different processes could be put in a line according to the sequence of operations required by the product (figure 4.7(a)). Such a layout is a production line and is termed a *product layout* since the system is geared to the production of a particular product. For example, on a car production line we might have a press followed by welding. Such a layout has a high initial capital cost, i.e. a high fixed cost, involved in setting up the production line for a particular product and relatively small operational costs, i.e. variable costs.

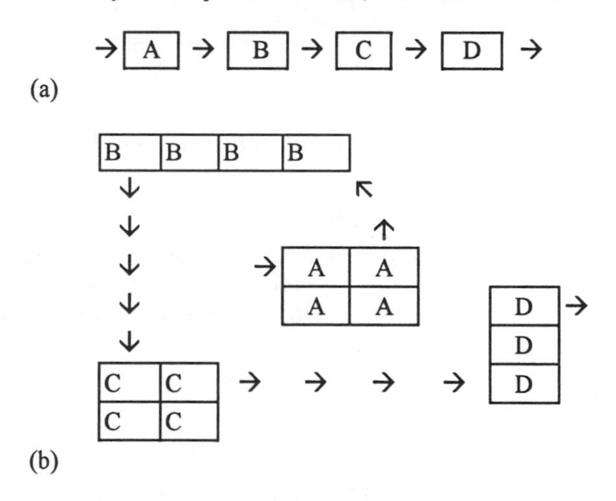

Figure 4.7 The routes followed by a product with (a) product layout, (b) process layout. A, B, C and D represent different types of machines.

Another possible layout is by *process*. Here all the machines concerned with a particular process are grouped together and the product moves to group A for one particular process and then to group B for the next process (figure 4.7(b)). Thus all the presses are grouped together. All the welding is grouped together. The route followed by different products can vary. This system is more flexible than product layout but more time is wasted in moving products between groups of machines. The capital costs, i.e. fixed costs, are much smaller than the product layout since the machines are not laid out for a specific product. The operational costs, i.e. the variable costs, are however much greater. The break-even charts for the two forms of layout might be of the form shown in figure 4.8. The data indicates that the process layout has a lower break-even point than the product layout. This means that profits can be made with smaller volumes with process layout than product layout. However, at large volumes when both forms of layout have passed their break-even points, the product layout gives greater profits than the process layout.

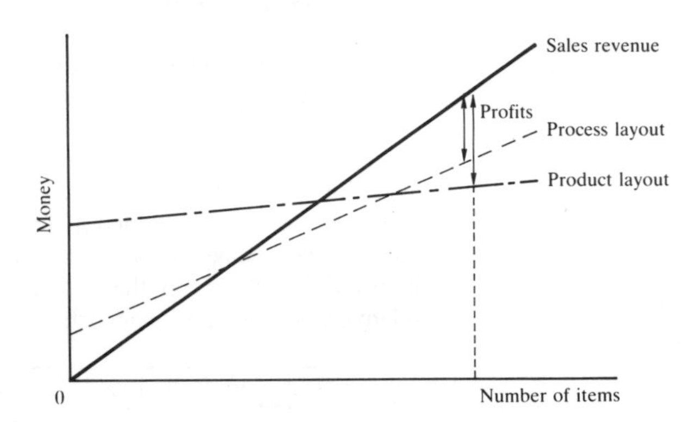

Figure 4.8 Break-even analysis of layouts

Example

Figure 4.9 shows a break-even chart with three different ways of achieving production. Which method (a) breaks even with the lowest number of items produced, (b) has the largest profit with 2000 items?

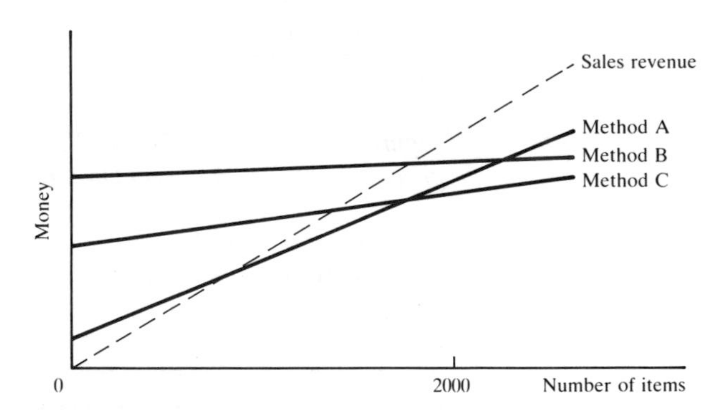

Figure 4.9 Example

(a) Break-even occurs at the smallest volume for method A.
(b) At 2000 items the greatest profit is made by method C.

4.5 Changes and break-even charts

The effect of changes on profits or losses can be shown by drawing lines on a break-even chart to represent the changes. Possible changes that might affect profits are:

1 There could be an increase in the volume of items produced and sold.
2 The selling price per item could be increased or decreased.
3 The fixed costs could increase or decrease.
4 The variable costs per item could increase or decrease.

To illustrate the effects of the above changes on break-even charts, an example is considered when each of the above items in turn is changed.

Consider, initially, that the basic costs and revenue for a particular product are: fixed cost £2000, variable cost £4 per item, selling price £8 per item. This data can be used to plot the break-even chart shown in figure 4.10 (this data is the same as used in a earlier example and figure 4.6).

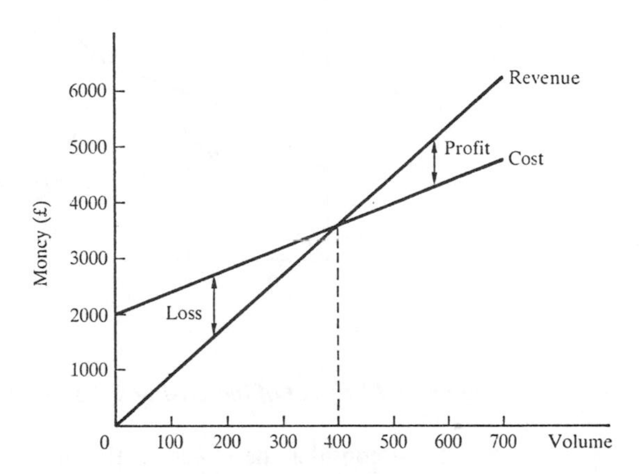

Figure 4.10 Initial break-even chart

Consider the effect of an increase in the volume of sales with no changes in the fixed cost, variable cost per item or selling price per item. As figure 4.10 indicates, above the break-even point the higher the volume the greater the profit and below the break-even point the greater the volume the less the losses. For example, with a volume of 100 items there is a loss of £1500. Increasing the volume to 200 items changes the loss to £1000. With a volume of 500 items there is a profit of £500. Increasing the volume to 600 items changes the profit to £1000 (see the example with figure 4.6).

Now consider what happens if the selling price is increased by £3 per item. This will result in the revenue changing to give:

Number of items	100	200	300	400	500	600
Initial revenue £	900	1800	2700	3600	4500	5400
New revenue £	1200	2400	3600	4800	6000	7200

If the new revenue data is plotted on the break-even chart (figure 4.11), we see that the result is a line for the revenue of greater slope, since the slope is the revenue per item. The effect of this is to give a lower break-even point, an increase in profit at sales volumes above the break-even point and a reduction in the losses at sales volumes below the break-even point.

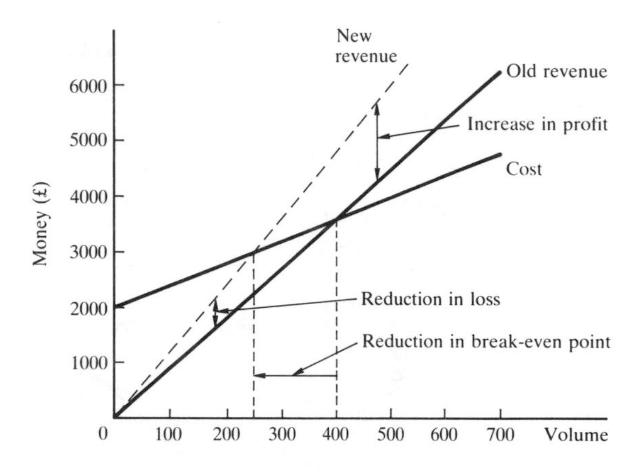

Figure 4.11 Effect of increasing sales price

Now consider the effect on the initial data and break-even chart of a decrease in fixed costs from £2000 to £1500. The variable costs and the selling price per item are considered to remain unchanged. The effect on the total cost of a number of items is, regardless of the number of items concerned, to decrease the total cost by £500. We thus have:

Number of items	100	200	300	400	500	600
Initial fixed cost £	2000	2000	2000	2000	2000	2000
Variable cost £	400	800	1200	1600	2000	2400
Initial total cost £	2400	2800	3200	3600	4000	4400
New fixed cost £	1500	1500	1500	1500	1500	1500
New total cost £	1900	2300	2700	3100	3500	3900

Figure 4.12 shows the new cost line plotted on the break-even chart. The new cost line is just parallel to the initial costs line, the difference between the lines being just the difference in fixed costs. The new line gives a reduction in the break-even volume, an increase in the profit at volumes above the break-even point and a decrease in the losses at volumes below the break-even point.

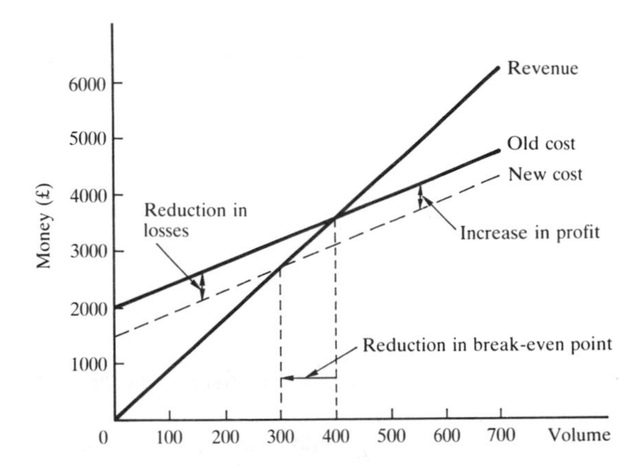

Figure 4.12 Effect of decreasing fixed costs

Now consider the effects on the initial data and break-even chart of an increase in the variable costs per item of £1 per item.

Number of items	100	200	300	400	500	600
Fixed cost £	2000	2000	2000	2000	2000	2000
Initial variable cost £	400	800	1200	1600	2000	2400
Initial total cost £	2400	2800	3200	3600	4000	4400
New variable cost £	500	1000	1500	2000	2500	3000
New total cost £	2500	3000	3500	4000	4500	5000

Figure 4.13 shows the new cost line plotted on the break-even chart. The new cost line has a different slope to the initial cost line. The result is an increase in the break-even point, an increase on losses at volumes below the break-even point and a decrease in profits at volumes above the break-even point.

Example

A company produces a product with fixed costs of £10 000, variable costs of £10 per item and a selling price of £20 per item. What changes

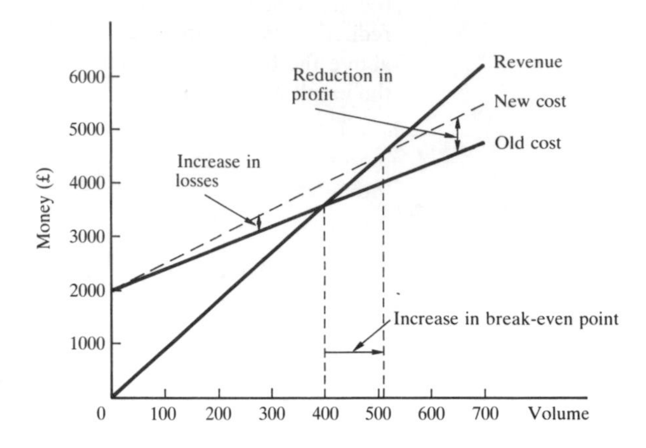

Figure 4.13 Effect of increasing variable costs

could the company make to reduce the break-even volume, i.e. the volume that has to be sold before a profit is made?

There are a number of possibilities:

1 Increasing the sale price per item will reduce the break-even volume (see figure 4.11).
2 Decreasing the fixed costs will reduce the break-even volume (see figure 4.12).
3 Decreasing the variable costs per item will reduce the break-even volume (the reverse of figure 4.13).

4.6 Make or buy?

The decision on whether a business should make or buy an item will be determined by a number of factors. These can include:

1 Can production make the item to the required quality, in the required volume and at the required time?
2 Are there suppliers who can make the item to the required quality, in the required volume and at the required time?
3 How does the cost of making the item compare with that of buying it?
4 Would making the item give greater existing plant and employee utilization, i.e. use capacity that would otherwise be idle?
5 Is the item of crucial importance so that if the supplier failed to deliver there would be serious problems?
6 Would the business be heavily dependent on just one supplier so that if he had problems the supply of the item could dry up?

Here, however, we will just consider the cost factor (item 3 above). If the items are to be bought-in then there are no fixed costs for the purchase of equipment to make them, all the cost being variable. With making them the business is likely to have some capital costs. Thus there will be both fixed costs and variable costs involved. Thus a graph of cost against number of items made for buying and for making might take the form shown in figure 4.14. At number of items *n* the costs of the two are the same. For numbers up to *n* it is cheaper to buy the items than make them. For numbers greater than *n* it is cheaper to make them than buy them. Thus the decision on whether to make or buy, if determined purely on cost grounds, will depend on the number of items required.

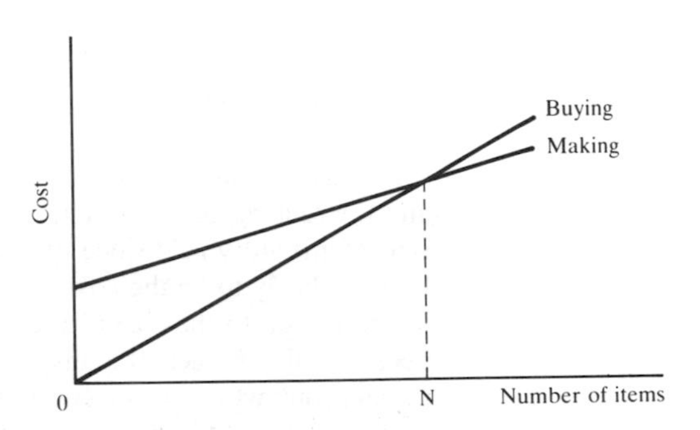

Figure 4.14 Buying or making

4.7 Inventory costs

The term *inventory* is used for items held in stock. This stock might be:

1 Materials and components which have been bought-in for use in the manufacture of products or in the administration of the business.
2 Finished products waiting orders from customers. Such stocks may enable fluctuating customer demand, e.g. seasonal demands, to be met by steady working over a longer period. For example, a business might make swim suits at the same rate over the entire year and store them in order to be able to cope with the large demand in the Spring and Summer. Producing them only when the demand is there would mean a much higher production capacity which might be idle for the remainder of the year. They can also enable the business to produce products in economic batches rather than just in, perhaps, the intermittent trickle that customer orders are received.
3 Part-processed products waiting for the next stage of production. Such work-in-progress stocks may be needed to adjust for the different rates of working between departments or perhaps stations along a production line. They can reduce delays due to defective work occurring at some stage in production and so, for example, avoid a production line coming to a halt when such an event occurs.

A business might thus hold stocks of raw materials, work-in-progress and finished goods. Such inventories cost money and are generally a major cost element. There are three main costs associated with inventories:

1. The cost of ordering stock. This includes the costs involved in the purchasing department ordering the stock, the costs of receiving the stock and inspecting it, etc.
2. The cost of holding the stock. This includes the rate of interest that could have been obtained from the capital if it had not been used to purchase the stock, the costs incurred for storing the stock, fire and insurance costs, the costs due to stock deteriorating, etc.
3. The cost of running out of stock. These are the costs that can be associated with the loss in production that occurs as a result of delays occurring from a lack of raw materials or no buffer stocks of work in progress, the loss of customer orders because goods cannot be delivered when required, etc.

The total inventory cost is the sum of the ordering cost, the holding cost and the running-out cost. The ordering cost will probably decrease with the size of inventory held since the larger an order placed for raw materials the lower is likely to be the cost per item. The holding cost is likely to increase in proportion to the size of the inventory. The running-out cost is likely to decrease the bigger the inventory held, there being less likelihood of running out when big inventories are held. Figure 4.15 shows how these costs might vary with the size of the inventory held. The total cost line is obtained by adding together the holding cost, ordering cost and running-out costs. The total costs line shown in the graph has a minimum at a particular inventory size. This is then the optimum inventory level.

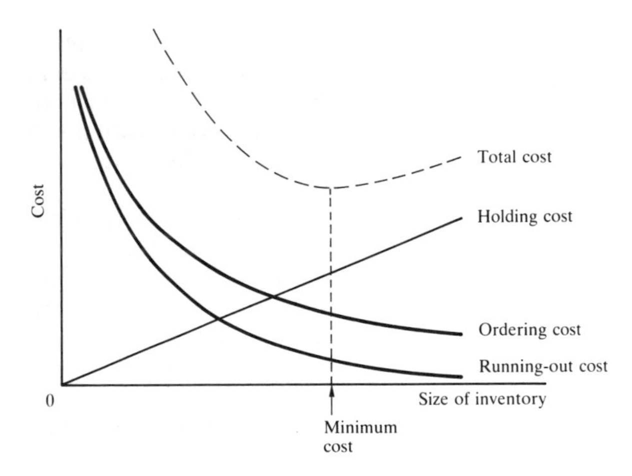

Figure 4.15 Inventory costs

Problems

Revision questions

1 Explain what is meant by capital costs and operational costs.

2 If the capital cost of an item is £20 000 and it is to be defrayed at a rate of 10% per year against production and typically 5000 items are produced per year, what will be the amount to be defrayed against each item?

3 A product has a fixed cost per year of £10 000 and a variable cost per item of £20. The selling price per item is £30. What will be the profit if there are sales of 2000 per year?

4 A product has a fixed cost per year of £10 000 and a variable cost per item of £20. What should the selling price per item be if a profit of £5000 is to be realized on the sale of 2000 items?

5 Explain what is meant by break-even chart.

6 A product has a fixed cost per year of £10 000 and a variable cost per item of £20. The selling price per item is £30. What will be the break-even volume?

7 A product has a fixed cost per year of £10 000 and a variable cost per item of £20. What should the selling price be if break-even is to occur with sales of 5000 items?

8 Sketch on a break-even chart the break-even graphs for two alternative methods of production when one has high fixed costs and low variable costs and the other has low fixed costs and high variable costs.

9 A product has fixed costs of £20 000, variable costs of £8 per item and a selling price of £12 per item.

(a) What is the total cost of producing 3000 items?
(b) What is the total cost of producing 5000 items?
(c) What is the profit/loss that would be made with sales of 4000 items?
(d) What is the profit/loss that would be made with sales of 9000 items?
(e) What is the break-even volume of sales?

10 A product has fixed costs of £150 000, variable costs of £20 per item and a selling price of £26 per item.

(a) What is the total cost of producing 15 000 items?
(b) What is the total cost of producing 30 000 items?
(c) What is the profit/loss that would be made with sales of 20 000 items?
(d) What is the profit/loss that would be made with sales of 40 000 items?

(e) What is the break-even volume of sales?

11 Explain why a business might carry stocks of raw materials, work-in-progress and finished goods and the costs associated with carrying such stocks.

Multiple choice questions
For problems 12 to 22, select from the answer options A, B, C or D the one correct answer.

12 The capital cost of owning a car are:

A The cost of buying the car
B The cost of buying petrol
C The cost of insurance
D The cost of maintenance

Problems 13-15 relate to the following information:

The costs of making 2000 items of a product are £4000 for the fixed costs and £6000 for the variable costs.

13 To make a profit of £2000, the selling price per item will have to be:

A £1
B £2
C £3
D £6

14 To break even, the selling price per item will have to be:

A £2
B £3
C £5
D £6

15 An increase in labour costs results in the variable costs being increased by 10 %. This will increase the selling price for break-even by:

A 5%
B 6%
C 10%
D 20%

16 Figure 4.16 shows the break-even chart for a product when manufactured by two different methods.

Decide whether each of these statements is True (T) or False (F).

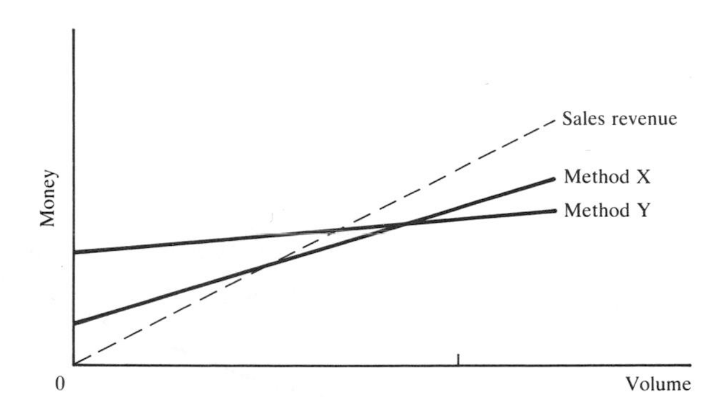

Figure 4.16 Problem 16

(i) Method X requires a smaller volume of product to be made before it becomes profitable.
(ii) Method Y is more profitable at volume V than method X.

Which option BEST describes the two statements?

A (i) T (ii) T
B (i) T (ii) F
C (i) F (ii) T
D (i) F (ii) F

17 For a break-even chart, the break-even volume is reduced if:

A The fixed costs are increased
B The variable cost per item is increased
C The selling price per item is increased
D The volume sold is increased

18 Figure 4.17 shows a break-even chart for a product.

Decide whether each of these statements is True (T) or False (F).

(i) The fixed costs of the product are £40 000
(ii) The profit with sales of 100 000 items is £60 000

Which option BEST describes the two statements?

A (i) T (ii) T
B (i) T (ii) F
C (i) F (ii) T
D (i) F (ii) F

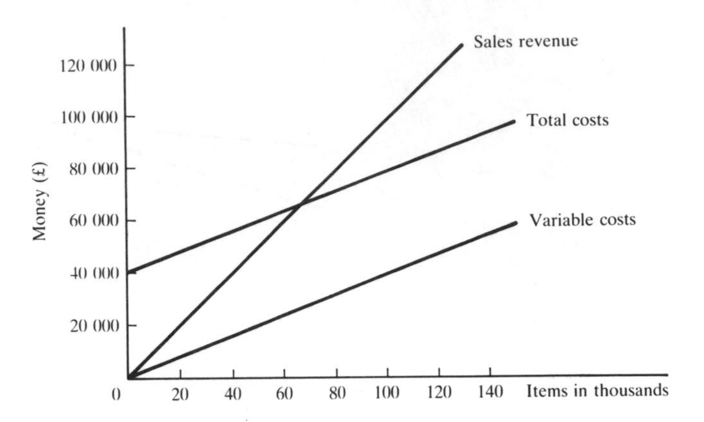

Figure 4.17 Problem 18

19 Figure 4.18 shows a break-even chart. Which is the dotted line showing an increase in the fixed costs?

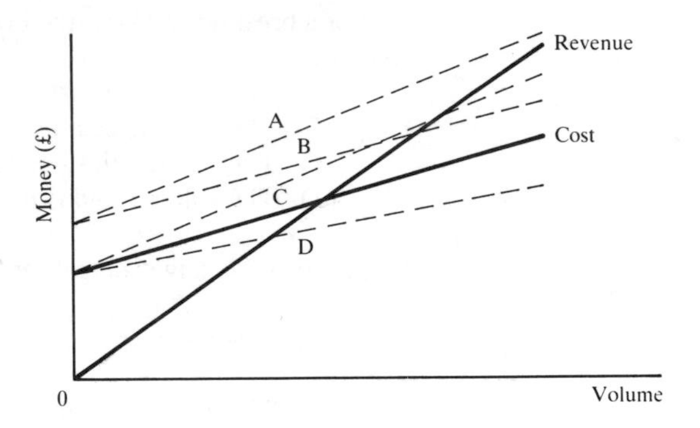

Figure 4.18 Problem 19

20 Figure 4.19 shows a break-even chart. Which is the dotted line showing an increase in the sale price per item?

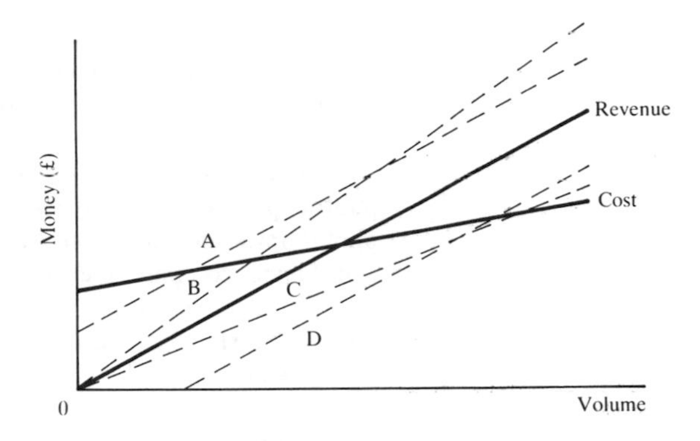

Figure 4.19 Problem 20

21 With a break-even chart for a product, the profit on the product is increased at a particular volume of sales above the break-even point if

A The selling price per item is increased
B The variable cost per item is increased
C The fixed costs are increased
D The volume at the break-even point is increased

22 Figure 4.20 shows how the ordering costs, the holding costs and the running-out costs for raw materials used in making a particular product depend on the size of the stocks held. Which of the points A, B, C or D best corresponds to the inventory size with the minimum cost?

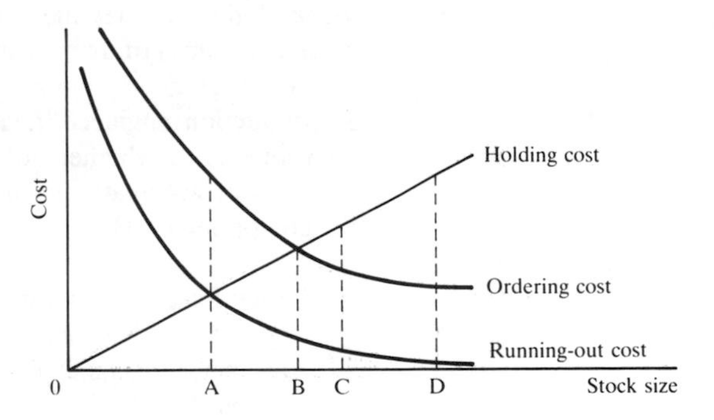

Figure 4.20 Problem 22

Assignments

23 Use data available in newspapers or magazines to arrive at the capital value of a particular model of a car after 1, 2, 3, 4 and 5 years. Hence establish the capital element of the costs which should be used to establish the running costs per mile/kilometre for each year.

24 Obtain data on the costs involved in having photocopies made at a photocopy shop or buying or leasing a photocopier and making photocopies in-house. Take into account the fixed and variable costs and hence compare the total costs involved. Hence determine the conditions under which it would be more economic to use the photocopy shop and those under which it would be better to use an in-house machine.

Case studies

25 The Modern Engineering Company makes car components. For a particular product the production manager is presented with a report from one of his engineers advocating that they replace the process currently used with automatic machinery. The data presented to the production manager for that operation is:

Number of items made per week 20 000
The selling price of each item is £0.15
Present manufacturing fixed cost per week is £1700 and the variable cost per item made is £0.05.
The automatic machine would increase the fixed costs to £1950 per week and reduce the variable costs to £0.03 per item.

(a) On the basis of the above data, would it be economic to install the automatic machinery?
(b) The production manager, however, feels that the data should be considered for other than just the present number of items made per week. Indicate what the effects would be of considering higher and smaller volumes of items being made.

26 A production engineer involved in production planning has to make decisions as to whether orders for turning should be routed to lathes, capstans or automatic machines. The following is the data he/she uses for comparing costs:

Lathe: fixed costs per item £17, variable costs per item £2.70
Capstan: fixed costs per item £27, variable costs per item £1.50
Automatic machine: fixed costs per item £170, variable costs per item £1.00

(a) Establish the order volumes for which each of the routes is the most economical method for turning.

(b) What would be the effect if a wages increase for the workers increased the variable costs by 10%?
(c) What would be the effect of new lathes being required and increasing their fixed costs by 10%?

27 The ABC Sports Company makes tennis rackets. Marketing as a result of a market survey estimate that a new model selling to shops at about £36 would have a potential sales of 20 000 per year. Product planning estimate the new equipment and a production line will be needed and that the capital cost will be £500 000. Of this capital cost it is estimated that £200 000 should be defrayed against the product in the first year of operation. In addition to this fixed cost there are other fixed costs for such items as the building and this will add a further £80 000. Variable costs on similar products lead to an estimate of variable cost per racket made of £22.

(a) Assuming the marketing information is correct, would the above data indicate that the racket should be made?
(b) It is considered that they ought to make a profit of 10% on each racket sold. Suggest how this could be achieved.
(c) The financial manager feels that inflation will result in the costs of materials rising and so the variable costs should be increased by 10%. What effect will this have on the above calculations?

28 ABC Electronics after doing well for a number of year was now finding it difficult to compete with some of the newer companies entering the market. Their costs had been cut to what was felt was the minimum of variable costs per item produced of £12 and fixed costs per year of £180 000. It was not considered possible to reduce these costs below this level. The production was running at the maximum capacity that could be realized of 120 000 items per year. Each item sold for £14. It was however felt that some action had to be taken in order to increase profits, otherwise the company would cease to trade. The management considerered that there were a number of possible solutions.

(a) One solution proposed was to raise the selling price by 10%. What effect might this have on profit if the sales continued at the same rate as before? What effect might this have if it resulted in the full capacity of operation not being used and it only be utilised at 90 % of full capacity?
(b) Another solution proposed was to increase the capacity by production staff working overtime. This could increase the number of items produced per year by 10% but would add £0.60 to the variable costs per item. What effect might this have on profit if all the items produced can be sold?
(c) In addition to increasing the capacity it was considered that they might have to cut prices by 10% in order to obtain the extra sales. What effect might this have on profits?

29 A business receives an order for one base for a special machine tool. This is not an item that the company already makes and so the method to be used has to be considered. There are two methods that production considers they could use. One is to make a grey iron casting of the base as a totality and the other to assemble it from steel sections and weld together the various parts after machining. For the casting, sand casting is proposed since only one item is required. For the casting the cost of making a sand pattern is £400, the variable costs of materials and labour are £590 and the fixed costs £500. The pattern could be used for making more than one mould. For the fabricated base, the variable costs associated with the machining, welding and raw materials is £670 and the fixed costs £300.

(a) For the one-off production, which method is most economical?
(b) Before the base can be made, marketing suggests that there might be a market for small numbers of such bases. Plot graphs showing how the costs vary with the number of items made and comment on the significance of the graphs in determining the optimum manufacturing method.

30 A company is considering two alternatives for the production of a hollow thermoplastic container. Injection moulding is estimated as having the following costs:

Fixed costs: installation £9000, die £2000
Variable costs per item: indirect labour £0.15, power £0.06
Direct labour per item £0.30
Direct materials per item £0.40
Finishing costs per item £1.00

The alternative, rotational moulding, is estimated as having the following costs:

Fixed costs: installation £1000, die £500
Variable costs per item: indirect labour £0.30, power £0.15
Direct labour per item £1.40
Direct materials per item £0.40
Finishing costs per item £0

(a) On the basis of the above costs, which process would be the most cost-effective if (i) 1000, (b) 100 000 items were required?
(b) The injection moulding process can produce 50 items per hour, the rotational moulding only 2 per hour. What implication might this have for using the equipment to produce a wider range of goods and so affect the decision regarding the choice of process?

5 Cost elements

5.1 Costs

This chapter is about the various elements that can be considered to make up the cost of making a product. By considering such elements, products can be costed and also such costing used as a management tool to monitor and control production. Chapter 6 is about monitoring and controlling costs.

To illustrate the types of cost elements that occur with manufacturing, consider the situation of a manufacturing business that makes, say, plastic tubing. The bills that will have to be paid might include:

Rent for the premises
Interest on bank loans
Stamps and stationery
Telephone bill
Payments for raw materials
Electricity for light, heat, and power
Maintenance items for production machines
Repairs to machines
New machines
Packaging for products
Delivery costs for products
Wages and salaries for those directly involved in production
Wages and salaries for those not directly involved in production
etc.

Some of these costs depend on the volume of output from the business. For example, the more plastic tubing they make the greater the cost for raw materials. Such costs are referred to as *variable costs* or *direct costs*. Figure 5.1(a) shows how such costs can vary directly in proportion to the volume of output. Some of the costs will not be affected by the volume of output. For example, the rent for the premises will be the same, regardless of whether any plastic tubing is produced or not. Such costs are referred to as *fixed costs*. Figure 5.1(b) shows how such costs are constant and do not vary as the output level changes. Some of the costs vary as the level of output changes but only indirectly. For example, maintenance costs for the machinery will increase the more the machinery is used. There might be planned maintenance which is undertaken whatever the output volume and so is a fixed cost and a variable element which is directly related to the output volume. The total cost thus increases indirectly rather than strictly in relation to the output level. Such a cost is referred to as an *indirect cost*. Figure 5.1(c) shows how such costs might vary with the output level. Such indirect costs are often broken down into the separate constituent fixed and variable cost elements and the elements then listed separately as fixed and variable costs.

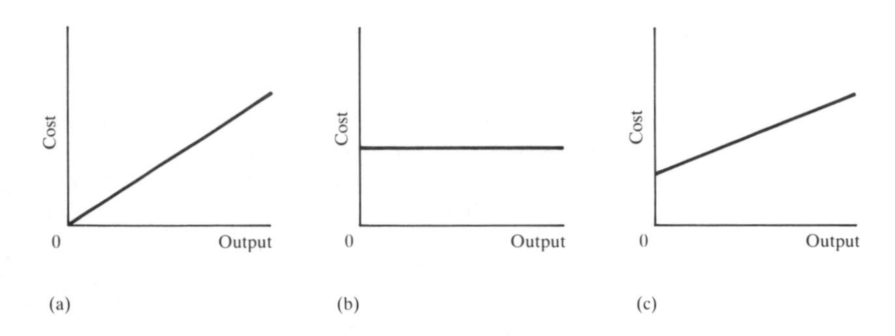

Figure 5.1 (a) Direct costs, (b) fixed costs, (c) indirect costs

The term *overheads* is used for those costs which are charges on the business and not charges directly related to the product. Thus the fixed costs and the indirect costs can be considered to be overhead costs. The indirect overhead costs are sometimes referred to as *variable overheads* since they vary when the output volume changes. Thus costs can be grouped as:

Direct costs or variable costs
Fixed overhead costs
Indirect overhead costs or variable overheads

The total costs of an organization for a particular output is found by adding together the direct costs, fixed overhead costs and indirect overhead costs for that level of output.

Total cost = direct costs + fixed costs + indirect costs

The profit made by a business is the difference between its total costs and the total sales revenue, i.e.

profit = sales revenue – total costs

When the sales revenue is greater than the total costs then the subtraction gives a positive answer and so there is a profit. If the total costs are greater than the sales revenue then the profit is negative, i.e. there is a loss. Such negative terms are usually indicated by putting brackets round them, e.g. (£3000), rather than writing them with a negative sign, i.e. – £3000.

Example

List the cost elements involved in buying and running a car, indicating which are variable, which fixed and which indirect costs.

The cost elements include:

Car purchase
Insurance
Petrol and oil
Road tax
Maintenance
Repairs
Parking fees
Parking fines
etc.

Variable costs which depend on the mileage covered are petrol and oil. Fixed costs are the purchase price, road tax, and insurance. Indirect costs are servicing, repairs, parking fees, and fines.

5.1.1 Historical and standard costing

The term *historical costing* is used when the costs are determined after the production of the goods has occurred. The term *standard costing* is used when costs are estimated before the production occurs. With historical costing, the costs are determined from records taken of the materials, labour and machines used during the production. With standard costing, the standards to be adopted for the costs of the various parts of the process are obtained from past experience of manufacturing the same or similar products. Thus standard costs are not the actual costs of production but estimates of what the costs will be. Such estimates enable cost control to be exercised over production in that there is a cost forecast before the production and this can be compared with the actual cost and the factors responsible for any discrepancy in costs identified.

5.2 Elements of manufacturing costs

The manufacturing cost of a product can be considered to consist of three basic cost elements: direct materials, direct labour and the overhead costs relevant to production, i.e.

total manufacturing cost of a product = direct materials cost
+ direct labour cost
+ overhead costs

The *direct materials cost* is the cost due to the materials which are directly used for making the product. Thus, for example, steel used to make a component is a direct material cost but materials used to maintain or repair the production machines used might not be. It might not be possible for the maintenance and repair materials to be identified with a particular product since they are used for the benefit of all the products produced by the machine. Such costs are then referred to as *indirect materials costs* and then included in the overhead costs.

The *direct labour cost* is the cost due to the labour directly used for making the product. This would thus include, for example, the wages of the operatives engaged in operating the machines making that particular product. It would not include the wages of workers who do not work directly on the product itself but assist in a general way with the manufacturing operation. The term *indirect labour cost* is used for labour that is only indirectly involved in the production. Such indirect labour is then included in the overhead costs. The sum of the direct labour and direct materials costs gives the total direct cost.

Total direct cost = direct materials cost + direct labour cost

The *overhead costs* are those costs related to manufacturing which are defrayed against the product. These would include indirect labour and materials costs, depreciation of machinery, rent of the factory, etc. Overhead costs are incurred for the benefit of all the products produced by a business and thus the amount of the overhead costs which should be defrayed against a particular product can only be an estimate. Methods of determining what element of the total overhead costs of a business to defray against a particular product are discussed later in this chapter.

For example, consider the production by a business of 10 000 items of a particular product. If the direct labour costs are £40 000, the direct materials costs £20 000 and the overhead costs which are defrayed against the product £20 000, then the total manufacturing cost of producing the 10 000 items is

Manufacturing cost elements	£
Direct labour	40 000
Direct materials	20 000
Overheads	20 000
Total	80 000

The cost per item is then £80 000/10 000 = £8.

Example

The ABC Machine Tool Company makes a single product. It produces 100 000 items of this product per year. The annual overhead bill is £400 000 and all the company overheads are defrayed against this single product. Direct materials costs are £2 per product item. The direct labour costs are £4 per product item. Maintenance costs are £50 000 per year, stores costs £30 000 and miscellaneous indirect overheads £50 000 per year. What is the cost per product item produced?

Since the company makes only one product, all the overhead costs are being defrayed against the one product. This term might include rent for the premises, insurance, etc. In addition we have the overheads due to production specific items. Thus we have:

Manufacturing cost elements per year	£
Direct labour for 100 000 items	200 000
Direct materials for 100 000 items	800 000
Overheads: maintenance	50 000
: stores	30 000
: miscellaneous	50 000
: general	400 000
Total	1 530 000

Thus the cost per item is £1 530 000/100 000 = £15.30.

5.3 Direct labour costs

With historical costing, the direct labour costs are determined from the records kept of each worker's involvement in the production process. Thus, for example, there might be two hours by an operator with machine X and then one hour by an operator of machine Y. The operator's time on machine X might be at the labour rate of £10 per hour and the operator's rate on machine Y at £12 per hour. The total cost is then

$$\text{total cost} = 2 \times 10 + 1 \times 12 = £32$$

With standard costing, the direct labour cost is found from considering the standard times specified for each part of the process. The *standard time* for a job is defined as the time needed for a qualified worker to carry out the job at a defined level of performance. Such times are determined by work-study staff timing workers carrying out the process concerned. The time to be used for a particular worker is then established by multiplying the standard time by a factor related to the performance of a particular worker when compared with that adopted for the standard. An operator rated as 100 is equivalent to the standard worker while one rated as less than 100 performs at a higher rate and one at more than 100 at a slower rate. Hence the standard cost for a job is

$$\text{standard cost} = \text{standard time} \times \text{operator rating} \times \text{labour rate}$$

Thus, for example, for a job for which the standard time is 2 hours, the operator rating 110 and the labour rate £12 per hour, the standard cost is

$$\text{standard cost} = 2 \times \frac{110}{100} \times 12 = £26.40$$

Example

Determine the direct labour cost for a job for which the standard time is 3 hours, the labour rate £12 per hour and the operator rating 90.

Using the above equation, then

$$\text{standard cost} = 3 \times \frac{90}{100} \times 12 = £32.40$$

Example

The ABC Engineering Company is preparing an estimate for the production of 120 000 pressings. Investigations indicate that, under standard conditions, a press can make 100 pressings per minute and requires one operator. The labour rate for press operators is £5 per hour. Estimate the standard direct labour cost for the 120 000 pressings.

The time needed to produce 120 000 pressings is

$$\text{time} = \frac{120\,000}{100} = 1200 \text{ minutes} = 20 \text{ hours}$$

Thus the cost is $20 \times 5 =$ £100.

5.4 Direct materials costs

Where materials are bought specifically for a particular job then the actual costs can be charged to that job. Thus, for example, we might have each unit of a product requiring one unit of material X, one unit of material Y and two units of material Z. If the cost of material X is £10 per unit, material Y £20 per unit and material Z £12 per unit, then

$$\text{direct materials cost per product unit} = 1 \times 10 + 1 \times 20 + 2 \times 12 = £54$$

In many instances, however, the materials used on a particular job are drawn from stocks held in the store by the company. Materials are thus bought for the store and then issued from the store for a particular job when requested by production. Since the materials drawn from the store for a particular job may use stock bought at different times and at different prices, it is not always feasible to use the actual costs of the materials in arriving at the direct materials cost. Methods that might be used in such circumstances are:

1 *Standard cost* The material issued from the store for a job is charged at a standard cost. This is a cost that has been chosen for that particular material and is not necessarily the real price paid for it.
2 *First in, first out (FIFO)* The cost of the oldest of that type of material in the store is used, it being assumed that the material that was first in is first out.
3 *Last in, first out (LIFO)* The cost of the most recent purchase of the material is used, it being assumed that the material last into the store is the first used.
4 *Highest in, first out* The cost is taken as being that of the most highly priced purchase, regardless of the date on which the material was bought.
5 *Average cost* The average price of the stock held of a particular material is used.
6 *Market price* The cost is taken as the market price of the material on the date the material is taken from the store, regardless of the price actually paid for it.

7 *Replacement pricing* The cost is taken as the price that it is anticipated will have to be paid to replace the material drawn from the store.

In general, the methods most used are standard costing where standard costing is used for the other aspects of the job and average cost in other situations.

Example

A batch of products requires 35 units of a particular material. This material is drawn from stock on February 12th. What will be the cost of the material using the (a) FIFO method, (b) LIFO method, (c) average cost method, given the following data for the materials in stock?

Opening stock bought January 1st: 20 units @ £10 each
Stock issued January 20th: 10 units
Stock purchased February 1st: 30 units @ £12 each

(a) The first in, first out (FIFO) method means that the oldest material in stock is used first. Thus the cost used is £10 per unit for the first 10 units. The next 25 units have to be drawn from the February 1st stock, these now being the first-in stock, and so are charged at £12 per unit. The total cost for the 35 units is thus $10 \times 10 + 24 \times 12 =$ £388.
(b) The last in, first out (LIFO) method means that the latest material in stock is used first. Thus the cost used is £12 per unit for the first 30 units. The next 5 units are drawn from the January 1st stock, these now being the last-in stock, and so are charged at £10 per unit. The total cost for the 305 units is thus $30 \times 12 + 5 \times 10 =$ £410.
(c) The average cost per unit is

$$\text{average} = \frac{20 \times 10 + 30 \times 12}{50} = £11.20 \text{ per unit}$$

Thus the cost of 30 units is $30 \times 11.20 =$ £336

5.5 Overhead costs

The term *overheads* is used to describe the costs which cannot be specifically allocated to any particular job or product but are indirectly part of the costs. Indirect materials are those materials which are used to further the manufacturing process but which cannot be directly identified in the end product, e.g. cutting oil or maintenance items for plant. Indirect labour consists of the wages and salaries paid to those not directly concerned with the production of the product, e.g. supervisors, managers, clerks, typists, salespeople, etc. Indirect expenses include all the expenses incurred by the business in carrying out its activities and which cannot directly be identified with a specific job or product, e.g. rent, electric power, insurance, etc.

Overhead costs can be classified as *fixed* if they remain constant and do not change as the output changes. Thus, for example, rent and insurance are fixed overhead costs. Overhead costs can be classified as *variable* if they are affected by the volume of output. Thus, for example, telephone charges have a fixed element which is the standing charge for connection to the telephone system and a variable element which is related to the number of telephone calls made. The number of calls made is likely to be related to the volume of output since the more output that is sold the greater the number of telephone calls that are likely to be made.

Table 5.5 gives some examples of fixed and variable overheads in relation to the production function, the sales function and general administration.

Table 5.1 Examples of overheads

Production overheads	Sales overheads	Administration overheads
Fixed		
Rent for factory	Sales office rent	Rent for office
Depreciation of plant	Depreciation of cars of representatives	Depreciation of office equipment
Insurance of stock and plant	Insurance of cars	Insurance of office equipment
Management salaries	Management salaries	Management salaries
Indirect materials	Catalogues	Stationery
Heating, lighting	Heating, lighting	Heating, lighting
Variable		
Indirect materials	Running costs for cars	Toner cartridges for printers/copiers
Supervisors' labour	Agents' commissions	Clerks' labour
Power		
Maintenance	Car maintenance/ repairs	Word processor maintenance/repairs

Overhead costs have to be recovered. Two methods by which overhead costs can be allocated against products, and hence form a cost element for each item sold, are discussed in the following sections. The methods are *absorption costing* and *marginal costing*.

Example

In running a car the owner incurs the following overhead charges. Which of them are fixed overheads and which variable?

Rent for the garage
Car insurance
Maintenance
Repairs

The rent for the garage and the car insurance do not depend on the number of miles/kilometres for which the car is used. They are thus fixed overhead charges. The maintenance and repair costs depend indirectly on the number of miles/kilometres for which the car is used. They are thus variable overheads.

5.6 Absorption costing

Absorption costing is the term used for the costing technique involving the absorption of all the overhead costs of the business into the cost of the products. Thus the overhead costs of the rent for the factory, the electrical power used, the telephone, etc. are all absorbed, together with all the other service costs such as the costs of the personnel department, sales staff, etc., into the cost of the products. All the costs of the business are absorbed by the cost of the products.

Consider a simple example to illustrate this. Mary and Ann run a market stall selling cups of coffee. Their sales and costs in a month are:

Rent for the stall	£400
Electricity	£ 30
Insurance	£ 50
Water standing charge	£ 20
Polystyrene cups	£ 12
Coffee	£ 20
Wages	£200
Number of cups sold	2250

The direct materials cost is £32 for the cups and the coffee, the direct labour charge £200 and

total overhead charge = 400 + 30 + 50 + 20 = £500

If all the costs are to be absorbed by the cost of the product then the total costs to be levied against the cost of cups of coffee are

total costs = 32 + 200 + 500 = £732

The cost to be levied against one cup of coffee is thus 732/2250 = £0.33. Here all the cost elements have been absorbed into the cost of the single product.

Example

A production department produces 1000 items of their product in a month. In that month the direct materials cost is £900, the direct labour cost £1200, and the production overheads £500. Non-manufacturing costs of £100 have to be levied against the product. What then is the total cost per item of that product?

The total production costs are the sum of the direct materials, the direct labour and the production overheads, i.e. 900 + 1200 + 500 = £2600. The cost to the business is the sum of the production cost and the non-manufacturing cost element. Thus the total cost levied against the product is.

total cost = 2600 + 100 = £2700

Thus the cost per item is 2700/1000 = £2.70.

5.6.1 Cost centres

The procedure adopted with absorption costing is to:

1. Divide the business into cost centres. The term *cost centre* is used for any section of the business for which costs are separately identified. Thus the production department might be a cost centre, or there might be a number of production cost centres according to the different products produced. The justification for a particular cost centre is that financial control is aided by separately identifying such costs. In a manufacturing business two types of cost centre can be identified, namely service and production cost centres. *Service cost centres* are those which do not directly make products, e.g. personnel, stores, etc. *Production cost centres* are those concerned with directly making products.
2. Apportion all the overhead costs between these cost centres. Thus, for example, the rent overhead cost might be apportioned to cost centres according to the floor areas they occupy. Lighting costs might be allocated according to the number of electric light fittings they have. Some overhead costs may be able to be directly attributed to just one cost centre, e.g. materials used in plant maintenance, while others are allocated to a number or all cost centres.
3. All the costs are then apportioned in some way to the product items or jobs and so absorbed and passed on in the total cost to the customer of the products.

The basis used for apportioning the overhead costs between cost centres is likely to depend on the overhead cost element concerned. Thus, for example, rent and insurance for the building might be apportioned on the basis of the floor area occupied by the cost centre concerned. Insurance for specific items might be apportioned according to the valuation of those items. Canteen costs might be apportioned according to the number of workers. Power, heat and light might be apportioned according to the floor area occupied. Indirect wages, such as those of general building maintenance, might be apportioned according to the amount of time they spend in a particular section.

Table 5.2 illustrates how some overhead costs might be allocated to cost centres, i.e. illustrating items 1 and 2 in the list above. Item 3 is discussed in the next section.

Table 5.2 Apportioning overhead costs

Overhead item	*Total cost*	*Service cost centres*			*Production cost centres*		
		Stores	*Main-tenance*	*Person-nel*	*A*	*B*	*C*
	£	£	£	£	£	£	£
Rent, allocated according to floor area	50 000	10 000	5 000	5 000	10 000	15 000	5 000
Lighting costs, allocated according to number of fittings	10 000	500	500	800	3 000	4 000	1 200
Heating costs, allocated according to cubic capacity	12 000	500	800	1 000	3 500	4 500	1 700
Insurance costs allocated according to valuation	2 000	400	100	300	300	500	400
Factory maintenance materials	2 000		2 000				

Example

The rent paid by a business for its factory is a fixed overhead of £20 000 per year. The business has three cost centres of administration, production and stores. Allocate the overhead to the three cost centres in accordance with the floor area each occupies. Administration has a floor area of 6000 m^2, production 10 000 m^2 and stores 4000 m^2.

The total factory floor area is 6000 + 10 000 + 4000 = 20 000 m^2. Thus the fraction of the floor area occupied by administration is 6000/20 000 and so the fraction of the overhead which is to be charged to the administration is (6000/20 000) × 20 000 = £6000. The fraction of the floor area occupied by production is 10 000/20 000 and so the overhead cost to be charged to it is (10 000/20 000) × 20 000 = £10 000. The fraction of the floor area occupied by stores cost centre is 4000/20 000 and so the overhead cost that is to be charged to the stores cost centre is (4000/20 000) × 20 000 = £4000.

5.6.2 Allocation of service department costs

Service department costs are not directly involved with the production process and thus their costs have to be passed, somehow, onto the buyers of the products by their costs being allocated to the producing cost centres. It is really only from the customers that a business can recoup its costs and thus, in the end, all the costs have to be paid by the customers. This allocation can be made in a number of ways. One way is to compute the manufacturing cost of an item or job by adding together its direct materials cost, direct labour cost and a production overhead element. Over the total number of items made by that production cost centre the overhead costs for that centre will be absorbed. The costs of the service departments might then be recouped by taking them into account by defraying them from the difference in sales price and production cost for the products. With this method we thus have a division of the total costs of the business as shown in figure 5.2.

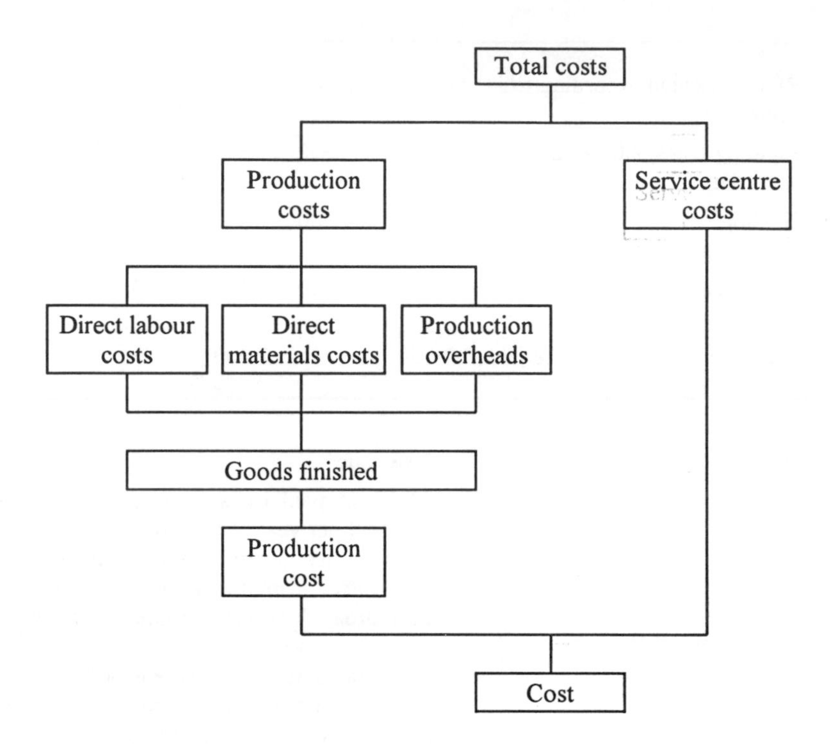

Figure 5.2. Cost allocation with absorption costing

An alternative is to divide the costs of the service departments between the production cost centres and include it as an addition to their overhead costs. Thus, for example, the costs of the stores can be divided between the production cost centres according to the fraction of the cost of direct materials each cost centre uses. Thus for the data quoted in table 5.2, the stores overhead cost of £11 400 might be divided so that production cost centre A is allocated £4000 of it, cost centre B £5000 of it and cost centre C £2400. Maintenance costs might be divided between the producing cost centres according to the number of labour hours each centre has. Thus for the data quoted in table 5.2, the maintenance overhead cost of £8400 might be divided so that production cost centre A is allocated £2000, cost centre B £2000 and cost centre C £4400. The end result of this allocation is that there is an overhead cost for each production cost centre and that the total of all these overhead costs is the total overhead cost of the business. All the costs are thus absorbed by the production cost centres. Figure 5.3 illustrates how the various cost elements combine in this way to give the product cost.

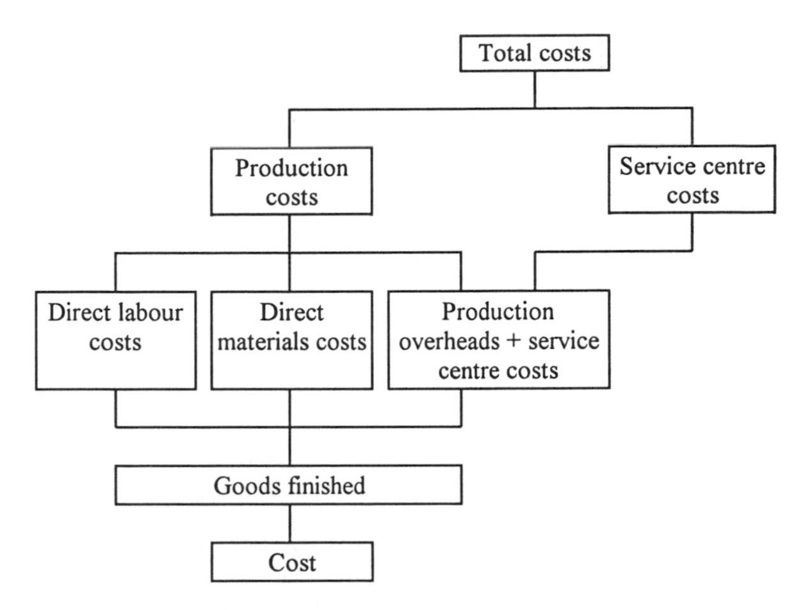

Figure 5.3 Cost allocation with absorption costing

Example

A manufacturing company has three production departments of machining, pressing and assembly and two service departments of engineering and administration. The overheads apportioned to each department are as given in table 5.3, all costs being in £. The engineering service and administration costs are to be apportioned between the production departments on the basis indicated in table 5.3. Hence apportioning the service department costs to the three production departments:

Table 5.3 Example

	Production departments			*Service departments*	
	Machining	*Pressing*	*Assembly*	*Engineer.*	*Admin.*
Overheads	150 000	65 000	50 000	45 000	30 000
% of eng. service costs	40	50	10		
% of admin. service costs	20	10	70		

For the machining department the overhead cost to be levied against their output is 150 000 + 40% of 45 000 + 20% of 30 000, i.e. a total of £174 000. For the pressing department the overhead cost to be levied against their output is 65 000 + 50% of 45 000 + 10% of 30 000, i.e. a total of £90 500. For the assembly department the overhead cost to be levied against their output is 50 000 + 10% of 45 000 + 70% of 35 000, i.e. a total of £79 000.

5.6.3 Allocation of overhead costs to products

The overhead costs for a particular production cost centre have to be, in some way, apportioned to the products produced at that cost centre. If the centre only produced one type of product then the overhead costs could be divided equally between the items. Thus if the overhead costs for a production centre are £50 000 and it produces 5000 items, then the overhead cost that has to be absorbed by an item of that product is 50 000/5000 = £10.

However, often a production centre will produce a variety of products and thus some way has to be used to apportion the overhead costs between the different products so that a cost element can be presented for each item produced. Methods that are commonly used are:

1. According to the direct labour costs. The greater the direct labour costs of a product the greater its share of the overheads.
2. According to the direct labour hours. The greater the number of direct labour hours needed for a product, the greater its share of the overhead.
3. According to the machine hours involved. The greater the number of machine hours needed to produce a product, the greater its share of the overhead.

The methods used for allocating such costs must be appropriate. Thus, for example, if a production department was apportioning overheads between three product cost centres then, if the products are labour intensive, it might apportion the costs on the basis of number of labour hours worked on each product. If the production is highly mechanized then they might be allocated on the basis of machine hours.

Consider, for example, a production department which has a total overhead cost of £10 000 per month and produces the following items in a month:

	Product A	Product B	Product C
Total direct labour costs	£12 000	£20 000	£15 000
Total direct labour hours	2400	4000	3000
Total machine hours	2000	3500	1000

When the overhead costs are apportioned according to the direct labour costs per product, then for product A we have

$$(12\ 000/47\ 000) \times 10\ 000 = £2553.19$$

For product B we have

$(20\ 000/47\ 000) \times 10\ 000 = £4255.32$

For product C we have

$(15\ 000/47\ 000) \times 10\ 000 = £3191.15$

When the overhead costs are apportioned according to the direct labour hours per product, then for product A we have

$(2400/9400) \times 10\ 000 = £2553.19.$

For product B we have

$(4000/9400) \times 10\ 000 = £\ 4255.32$

For product C we have

$(3000/9400) \times 10\ 000 = £3191.49$

When the overhead costs are apportioned according to the number of machine hours per product, then for product A we have

$(2000/6500) \times 10\ 000 = £3076.92$

For product B we have

$(3500/6500) \times 10\ 000 = £5384.62$

For product C we have

$(1000/6500) \times 10\ 000 = £1538.46$

Example

The production department of the XYZ Engineering Company makes 4000 items of its product X per month. This involves:

50 000 machine hours,
27 000 direct labour hours,
£100 000 direct labour costs

The production department also produces products Y and Z and these together account for

70 000 machine hours
40 000 direct labour hours
£110 000 direct labour costs

If the overhead costs that have to be absorbed are £85 000, what will be the cost that has to be absorbed in each item of product X if the costs are apportioned on the basis of (a) number of machine hours, (b) direct labour hours, (c) direct labour costs?

(a) Product X takes 50 000 machine hours out of a total of 50 000 + 70 000 = 120 000 machine hours. Thus the fraction of the overhead costs to be allocated to product X is (50 000/120 000) and so is a cost of

$$\text{cost} = (50\,000/120\,000) \times 100\,000 = £35\,417$$

This is a cost per item of product of 35 417/4000 = £8.85.

(b) Product X takes 27 000 direct labour hours out of a total labour hours of 27 000 + 40 000 = 67 000 direct labour hours. Thus the fraction of the overhead costs to be allocated to product X is (27 000/67 000) and so is a cost of

$$\text{cost} = (27\,000/67\,000) \times 100\,000 = £34\,354$$

This is a cost per item of product of 34 354/4000 = £8.56.

(c) Product X takes £100 000 of direct labour costs out of a total of 100 000 + 110 000 = £210 000 direct labour costs. Thus the fraction of the overhead costs to be allocated to product X is (100 000/210 000) and so is a cost of

$$\text{cost} = (100\,000/210\,000) \times 100\,000 = £40\,476$$

This is a cost per item of product X of 40 476/4000 = £10.12.

5.7 Marginal costing

With absorption costing all the overheads costs, both fixed and variable, are allocated in some way to the various products made by a business. The direct costs of production can easily be allocated to a particular product, but with overheads the problem is more difficult and some rule has to be devised as to how to apportion the overhead costs. The rules used are quite arbitary. Fixed overhead costs do not vary with output and are incurred in respect to a definite period of time, e.g. rent per year, while variable overhead costs are related to the output. There is thus an argument that, in considering the viability of making a product, only the variable overhead costs should be allocated to a product. The difference between the cost arrived at in this way and the selling price gives a return, called the *contribution*, which has thus to cover both the fixed overheads and the profit. Thus we have the equations:

$$\text{Contribution} = \text{sales revenue} - (\text{direct costs} + \text{variable overheads})$$

$$\text{Profit} = \text{contribution} - \text{fixed overhead costs}$$

This method of costing is known as marginal costing. Figure 5.4 illustrates these points.

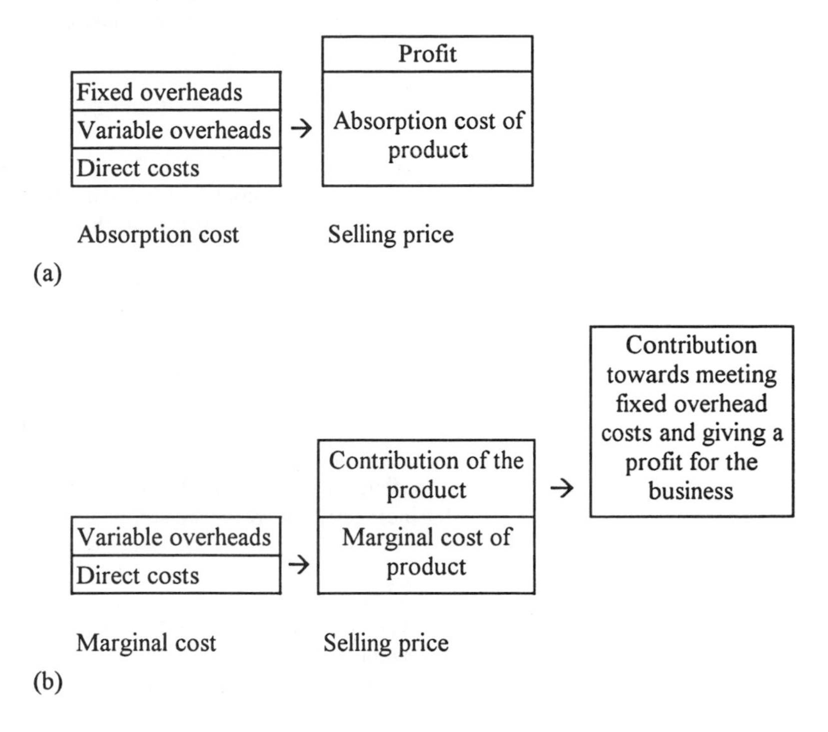

Figure 5.4 (a) Absorption costing, (b) marginal costing

The following example illustrates the differences between absorption and marginal costing. Consider a product for which 1000 items are produced and which has the following costs:

Direct materials costs	£2000
Direct labour costs	£1200
Variable overhead costs	£1200
Fixed overhead costs	£3800

The fixed overhead costs, such as factory rent, were allocated to the product on the basis of the floor area devoted to that product.

With absorption costing, the total costs to be levied against the product are the sum of the direct materials costs, the direct labour costs and the total overhead costs. Thus the cost of the 1000 items is

$$\text{cost} = 2000 + 1200 + 1200 + 3800 = £7200$$

A profit element of £800 might then be added to give a total selling price of £8000 for the items.

With marginal costing, the total costs to be levied against the product are the sum of the direct materials costs, the direct labour costs and the variable overhead costs. Thus the cost of the 1000 items is

$$\text{cost} = 2000 + 1200 + 1200 = £4400$$

If the selling price is fixed as £8000 then the product makes a contribution of £3600. This contribution can then be used, together with the contributions from the other products of the company, to meet the fixed overhead costs and give a profit to the company.

Marginal costing can simplify the costing of products in that only the costs directly related to those incurred in the production of the product are considered. It also has the advantage that it identifies the contribution a product makes and decisions can be made on that basis as to whether to make a product or continue with a product. Management can decide, for example, to set the price of a product so that it perhaps makes only a small contribution and balance this by some other product making a bigger contribution. The sum of the contributions from all the products of a business must, however, be sufficient to cover the costs and give a profit.

Example

The ABC Engineering Company is a small company which makes a single product for which the following are their estimated product-related costs per month:

Direct materials	£4000
Direct labour	£7000
Maintenance	£2000
Stores	£1000
Other variable overheads	£2000

The company has fixed overhead costs of £12 000 per year. Sales are estimated as being 60 000 items per year at a price of £4 per item. What is the contribution made by the product? What is the profit made per month by the company?

The marginal cost is the sum of the direct costs and the variable overhead costs. It is thus, per month,

$$\text{marginal cost} = 4000 + 7000 + 2000 + 1000 + 2000 = £16\,000$$

The contribution is the difference between the marginal cost and the sales revenue. The average number of items to be sold per month is 5000 and the total sales revenue per month is thus estimated as £20 000. Thus the contribution made per month is £4000.

Since the company has only a single product, this contribution has to meet the fixed costs and provide the profit. The fixed costs per month are £1000. Thus the company is making a profit of £3000 per month.

Problems

Revision questions

1 Explain the terms (a) direct costs, (b) indirect costs, (c) fixed costs and give an example of each.

2 Explain the terms (a) direct materials costs, (b) direct labour costs, (c) manufacturing overhead costs and give an example of each.

3 Explain what is meant by absorption costing.

4 Explain the term cost centre.

5 A company makes 1000 items of product X in a week. The direct materials cost of that production is £2000 and the direct labour costs £5000. If the overhead costs to be absorbed by that product's costs are £1000, what will be the total cost per item of product?

6 A company makes 500 items of product X and 200 items of product Y per week. The machine hours involved in the production are 1000 hours for product X and 600 for product Y. The overhead costs to be absorbed by the production of these two items are £1200. What will be the overhead cost to be absorbed by each item of product X and each item of product Y if the costs are apportioned according to the number of machine hours involved?

7 The rent paid by a business for its factory is a fixed overhead of £50 000 per year. The business has three cost centres of administration, production and stores. Allocate the overhead to the three cost centres in accordance with the floor area each occupies. Administration has a floor area of 7000 m^2, production 14 000 m^2 and stores 4000 m^2.

8 A production department has an overhead cost of £20 000 for the rent of the premises allocated to it. The production department has three cost centres, one for each of the products made, i.e. product X, product Y and product Z. There are 50 people employed in producing product X, 20 in producing product Y and 30 in producing product Z. Allocate the overhead cost to the three cost centres on the basis of the number of people employed at each.

9 Explain how absorption and marginal costing differ. Illustrate your answer by giving a simple example involving both methods of costing.

10 A company makes a product which has direct materials costs of £3000, direct labour costs of £5000, variable overhead costs of £4000 and

fixed overhead costs of £2000. The sales revenue from the product is £16 000. What will be the contribution made by the product?

11 On January 1st stores received 10 units of a material at a cost of £15 each unit. On February 1st they received 10 units of the material at a cost of £17 each unit. On February 10th, production requisitioned 8 units of the material. What is the cost to production of the units if they are costed on the basis of (a) the first in, first out method, (b) the last in, first out method, (c) the average method?

12 Explain what is meant by standard costing.

Multiple choice questions
For problems 13 to 25, select from the answer options A, B, C or D the one correct answer.

13 A machine with its operator is used to turn components. The operators time is charged at the rate of £10 per hour. The operator, in a particular week spends 30 hours on making 100 items of product X and 10 hours on 20 items of product Y. The materials involved in making product X cost £5 per item and for product B £3 per item. The maintenance costs for the machine per week are £30 of which £10 is materials costs and £20 labour costs.

Decide whether each of these statements is True (T) or False (F).
(i) The direct labour cost per item for product X is £3.
(ii) The direct materials cost per item for product X is £5.

Which option BEST describes the two statements?

A (i) T (ii) T
B (i) T (ii) F
C (i) F (ii) T
D (i) F (ii) F

Questions 14-15 relate to the following information:

The XYZ Electronics Company manufactures two products, P and Q. In a week it manufactures 100 items of its product P and 40 items of product Q. For product P the number of machine hours in the production department is 5 hours per item and for product B 2 hours per item. Product P requires a total labour time of 10 hours per item and product Q 5 hours per item. The labour is costed at an average rate of £5 per hour. The direct materials costs of product P are £4 per item and for product Q £10 per item. The overheads costs per week are £800.

14 When the overhead costs are allocated on the basis of machine hours, the cost to be allocated to an item of product P will be:

A £1.10
B £5.71
C £6.90
D £8.00

15 When the overhead costs are allotted on the basis of labour costs, the cost to be allocated to an item of product P will be:

A £1.33
B £4.17
C £6.67
D £8.00

16 The total direct costs, to the nearest pound, of an item of product P will be:

A £35
B £54
C £60
D £67

17 Decide whether each of the statements is True (T) or False (F).
(i) With absorption costing, all the overhead costs of a business are used to arrive at the costs of products.
(ii) With marginal costing, the overhead costs of a business are not used to arrive at the costs of products.

Which option BEST describes the two statements?

A (i) T (ii) T
B (i) T (ii) F
C (i) F (ii) T
D (i) F (ii) F

Questions 18-19 relate to the following information:

The following are cost elements which can figure in the total cost of a product:

(i) Direct materials costs
(ii) Direct labour costs
(iii) Fixed overhead costs
(iv) Variable overhead costs

18 The costs elements which will figure in the direct cost of a product are:

A (i) only
B (i) and (ii) only
C (iii) and (iv) only

D (i), (ii), (iii) and (iv)

19 The costs which will figure in the cost of production of a product are:

A (i) only
B (i) and (ii) only
C (iii) and (iv) only
D (i), (ii), (iii) and (iv)

20 Decide whether each of the statements is True (T) or False (F).

(i) Marginal costing only involves taking account of the direct materials and direct labour costs.
(ii) With marginal costing the term contribution is used for the amount of profit given by a product.

Which option BEST describes the two statements?

A (i) T (ii) T
B (i) T (ii) F
C (i) F (ii) T
D (i) F (ii) F

21 Which one of the following is a direct labour cost for production costs for a product?

A Salaries of salesmen/women obtaining orders for the product
B Wages of machine operators working on the product
C Work manager's salary
D Depreciation of machines used for making the product

22 Which one of the following is an indirect materials cost for production costs for a product?

A Materials used for general maintenance of the machines used
B Raw materials used in the manufacture of the product
C Components bought for assembly to make the product
D Components bought for incorporation in the product

Questions 23-25 relate to the following information:

A company starts on January 1st with a stock of a particular material obtained by a purchase of 20 units at a cost of £14 per unit. On January 20th 40 units of the material are purchased at a cost of £15 per unit. On January 30th, production withdraws 40 units of the material from stock.

23 The cost of the 40 units for production when reckoned on the basis of first in, first out is

A £560
B £580
C £587
D £600

24 The cost of the 40 units for production when reckoned on the basis of last in, first out is

A £560
B £580
C £587
D £600

25 The cost of the 40 units for production when reckoned on the basis of average cost is

A £560
B £580
C £587
D £600

Assignments

26 Identify the cost centre in which you study or work. List the main direct costs which you think the cost centre will be responsible for and the elements which might be part of the overhead costs of the college/school or business concerned.

27 A college operates a stationery shop where students can purchase stationery. The shop purchases the stationery in anticipation of sales to the students, maintaining a stock from which sales are made. Purchases, because they may not all be made at one time, may be at a variety of prices. Propose a costing method for the stationery. If you are working in a college with such a shop you could investigate how they cost the stationery.

Case studies

28 The ABC Gearboxes Company manufactures components for the car industry. They propose submitting a tender for 1000 gearboxes required by a car manufacturer. The costing section is asked to prepare an estimate of the costs that would be involved. On the basis of previous experience of producing similar gearboxes they estimate that the direct materials costs will be £20 000. The standard times for the production of such gearboxes are estimated as being 10 hours per gearbox. The average rate for the labour involved is £5 per hour. The tooling costs for the gearboxes is estimated, on the basis of previous experience, to be £6000. The overhead costs involved are, fixed overheads £14 000 and variable overheads £4000.

(a) Estimate the cost of a gearbox.
(b) It is considered that by the time the order is received there will have been a wage settlement which will increase the average labour costs by 10%. Give a revised estimate of the cost of a gearbox, taking this increase into account.
(c) Some time later the company is asked to submit a tender for a further 500 gearboxes. The tooling for the previous job can be used again. If the variable overhead costs are assumed to vary in proportion to the number of gearboxes, estimate the cost of a gearbox for this order.

29 PQR Engineering Company is a manufacturing organization with two production departments and two service departments. One of the production departments is concerned with machining and the other with assembly. The service departments are administration, finance and stores. The organization manufactures small batches of components, against orders, for car manufacturers. Each component involves both the production departments.

Devise a costing system which could be used to cost products. Your system should indicate:
(a) what cost centres are to be used and the reasons for their selection,
(b) how overheads costs are to be met,
(c) how an estimate of the cost of a product is to be obtained.

30 The marketing department of the Alpha, Beta, Gamma Engineering Company has researched the market and come up with proposals for two products, alpha and beta. The following is the information derived for the two products.

Product alpha would incur incurred a tooling cost of £200 000. The market is estimated to be 5000 items per year for five years. In view of competing products, marketing envisage a selling price of £50. The direct costs per item are:

Direct materials £5
Direct labour £10

Fixed overheads per item are allocated on the basis of direct labour hours at £10 per hour. Product alpha requires 2 hours of direct labour for each item.

Product beta would incur a tooling cost of £100 000. The market is estimated to be 4000 items per year for five years. In view of competing products, marketing envisage a selling price of £20. The direct costs per item are:

Direct materials £7.50
Direct labour £2.50

Fixed overheads per item are allocated on the basis of direct labour hours at £10 per hour. Product beta requires 1 hour of direct labour for each item.

(a) The company can only produce one of the above products. Which product would lead to the most profit?
(b) How would the costs, and hence the profits, be affected if product alpha could have a production run lasting 8 years?

31 The XYZ Car Components Company makes three components, X, Y and Z, for the car manufacturing industry. There are three production cost centres, one for casting, one for machining and one for assembly and packing. Not all the products pass through each department. The following are the direct and variable overhead costs per month that have been estimated for each of the products.

Product X
Casting department
Direct labour £3000, Direct materials £2000
Variable overheads: Stores £3000, Maintenance £1000
Assembly and packing
Direct labour £1000, Direct materials £1000
Variable overheads: Stores £500, Maintenance £500

Product Y
Casting department
Direct labour £3500, Direct materials £4000
Variable overheads: Stores £3500, Maintenance £1000
Machining department
Direct labour £2000, Direct materials £500
Variable overheads: Stores £500, Maintenance £1000
Assembly and packing department
Direct labour £1200, Direct materials £500
Variable overheads: Stores £600, Maintenance £500

Product Z
Machining department
Direct labour £5000, Direct materials £4000
Variable overheads: Stores £3000, Maintenance £1500
Assembly and packing department
Direct labour £2000, Direct materials £2000
Variable overheads: Stores £1000, Maintenance £500

The fixed overheads for the company are £20 000 per year. The sales revenue that is to be expected per month from each of the three products is:

Product X £25 000
Product Y £35 000

Product Z £20 000

(a) Use marginal costing to estimate the contribution to be made by each product each month.
(b) Which product contributes least and may need reconsideration as to its viability?
(c) Estimate the total profit to be made per month by the company.

32 The Modern Electronics Company plans to make a product for which marketing estimates that sales per year are likely to be no less than 2000 and might rise as high as 14 000. The sales price used for the estimates is £3 per item. The direct labour and materials costs per item amount to £1.50 and the variable overhead costs to £0.50 per item.

(a) What will be the contribution made at the estimated minimum and maximum sales volumes?
(b) The Company consider it will be uneconomic to produce the item if the contribution to the fixed overheads is less than £10 000. What will be the volume of sales required for the company to break even?
(c) If the company would like to make a profit of 10% on each item sold, what will be the contribution at the estimated minimum and maximum sales volumes towards the fixed overhead costs?

33 The ABC Engineering Company has been asked to submit a quote for a one-off product. The costing section have estimated the costs to be:

Direct labour 800 hours at an average cost of £5 per hour
Direct materials £600
Special tooling £500

The special tooling is not expected to be of any use after the product has been made. The company adopts a policy of absorbing fixed overhead costs at £2 per hour of direct labour and variable overhead costs at £1 per hour of direct labour.

(a) Estimate the cost of the product using absorption costing.
(b) Estimate the marginal cost of the product.
(c) Because the company is not running at full capacity the management take the decision that they would not try to cover all the fixed costs, as long as there was some contribution to those costs. Propose a selling price for the product.

34 The Alpha Engineering Company makes a single product for which they already have an established demand during the next year for 2000 items at a selling price of £12 per item. The costs incurred for this product and other costs are:

Direct labour £7500
Direct materials £5500

Production overheads: fixed £500
Production overheads: variable £4500
Administration expenses £1000
Sales and distribution expenses £800
Accounting expenses £200

(a) The Company uses marginal costing. What will be the contribution made by the product?
(b) During the year the company is approached by a new customer who would take 100 units if the price was £9 per item. The only extra costs that the company would incur are the variable costs. Should the company accept the order?

6 Budgetary control

6.1 Budgets

This chapter, and the next, are concerned with the planning, monitoring and control of costs in a business. *Planning* involves considering what the future will hold for a business, making forecasts about the future, and then arriving at decisions on what the business should do in the future and the methods to be used to obtain the required results. *Monitoring* involves checking how performance towards those results is occurring. *Control* is the measurement and correcting of the actual performance to ensure that the chosen course of action occurs.

This chapter is concerned with budgets. The term *budget* is used for a quantitative economic plan for a business in respect of a period of time. It outlines what a business wants to achieve, what it needs to do to obtain those results and how much it will cost. The budget is quantitative, i.e. expressed in numbers, and is expressed in such terms as numbers of product items to be sold, numbers of labour hours to be expended etc, and amounts of money to come in as income or go out as costs. They usually show the planned income to be generated and expenditure to be incurred during a period of time and the capital to be employed to attain the results aimed for by the business. A budget is a plan of what the business is aiming to achieve.

We need to distinguish between forecasts and plans. A forecast is what is expected to happen, a plan is what you aim to make happen. Budgets are not forecasts but plans. They are, however, likely to be based on forecasts. A budget is always for some period of time. This might be for a month, a quarter, a year, or perhaps five-years.

6.1.1 Functions of budgets

The functions of budgets can be considered to include:

1 Refining the aims of an organization into detailed plans to be implemented. The budgeting process involves managers in planning for future operations, making forecasts, considering how conditions might change and how they might respond. It then indicates how they aim to proceed to achieve the aims of the organization by giving details of the plans.
2 Co-ordinating the activities of the various parts of an organization. Budgets serve as vehicles by which the various parts of a business can bring together their plans to be co-ordinated into an overall plan for the business. For example, sales and production have to co-ordinate their plans as to the products, and numbers of items of each product, to be produced and sold.
3 Communicating plans to all parts of a business. Through a budget, management are able to communicate their expectations to all parts of a

business. It informs employees just what they are expected to do and what the other parts of the business will be doing.

4 Motivating managers. Budgets set targets and in doing so are a useful method for influencing the behaviour of managers, motivating them to achieve the goals set out.
5 Acting as a control for a business. Actual results can be compared with the budgeted amounts and attention thus directed to those areas where the performance does not come up to that planned.

6.2 Types of budgets

Budgets enable control to be exercised, i.e. targets are set and the actual results are compared with them. Thus there is likely to be a budget for each type of function, for each division, and for each type of cost. In this section we consider a number of budgets:

1 Sales budget
2 Production budget
3 Direct materials and materials purchasing budgets
4 Direct labour budget
5 Overheads budget
6 Cash budget
7 Master budget

The sales budget is a key budget from which the production budget, and consequently the materials and labour budgets, emanate. There is, however, the point that it is no use the sales budget listing sales of items beyond the capability of production to produce.

6.2.1 Sales budget

The *sales budget* shows the quantities of each product which the business plans to sell and the intended selling price. It thus predicts the total revenue that the business will receive from sales of its products over some period of time. A common length of time used is one year. It might thus look like:

Sales budget for year

Products	Items sold	Selling price £	Total revenue £
X	6 000	100	600 000
Y	3 000	150	450 000
Z	2 000	120	240 000
			1 290 000

Such budgets are, however, also likely to be broken down to give details for each month or quarter. Also, the sales budget might be broken down into other sales budgets, e.g. for each sales territory. Thus the budget for the sales in the North of the country might be:

North territory sales budget for year

Products	Items sold	Selling price £	Total revenue £
X	1 000	100	100 000
Y	600	150	90 000
Z	200	100	20 000
			210 000

Example

The ABC Manufacturing Company sales manager has obtained forecasts of sales for the next year for its products of:

Product A 1000 items
Product B 2000 items
Product C 3000 items

The sales prices that it is envisaged will be used are:

Product A £5 per item
Product B £4 per item
Product C £4 per item

Production indicates that these quantities can be produced in the year. Hence prepare the sales budget for the year.

The budget might thus look like:

Products	Items sold	Selling price £	Total revenue £
A	1000	5	5 000
B	2000	4	8 000
C	3000	4	12 000
			25 000

6.2.2 Production budget

The sales budget provides the basic data for constructing other budgets, such as the production budgets, administration budgets, distribution budgets, etc. The *production budget* is concerned with the plans for what products and in what quantities are required to be produced in some period of time, e.g. a year. Existing stocks of products and stocks required to be maintained have to be taken into account. Thus we might have, arising from the sales budget given in section 6.2.1:

Production budget for year

	Product X	Product Y	Product Z
Items to be sold	6 000	3 000	2 000
Planned closing stock at end of period	500	300	100
Total items required for sales and stock	6 500	3 300	2 100
Less planned opening stock	200	400	100
Items to be produced	6 300	2 900	2 000

In the above budget, the number of items to be sold is obtained from the sales budget. The above is the budget for the year. Arising from that, production will need to establish the budgets for perhaps each quarter, month, week. The questions that then have to be posed are:

1. Should production plan to obtain an even flow of each product throughout the year?
2. Would it be a more economic use of resources to produce the products in batches?
3. Is there a requirement for stocks of the product to be kept at the same level for each month of the year or are there perhaps some seasonal factors which mean that higher stocks should be available in certain months?

Consider product X in the above production budget and suppose that the demand for the product is largely seasonal, greater quantities being required in Spring. Suppose production can manage, when working at full capacity, 2000 items per month. We might then have a budget for product X of the form:

Production budget for product X

	Jan.	Feb.	Mar.	Apr.
Opening stock	200	1400	2400	1400
Items produced	2000	2000	2000	300
	2200	3400	4400	1700
Less sales	800	1000	3000	500
Closing stock	1400	2400	1400	1200

Thus for the month of January, there is an opening stock of 200 items and production is required to work at full capacity and produce 2000 items of product X. This results in a closing stock at the end of the month of 1400 items. Working again at full capacity in February, the sales of 1000 items can be met and there is a closing stock of 2400 items. This high closing stock is necessary if there are to be sufficient items to meet the March sales with production working at full capacity. The entire production for the year of product X has been concentrated into the first few months of the year. The stock remaining after April is enough to cover the rest of the sales in the year and leave the required stock level at the end of the year.

Example

The Alpha Electronics Company requires production of the product alpha to be maintained at a steady rate for the quarter. The sales budget for the quarter envisages some variations in sales of items of the product per month, these projections being:

Jan.	Feb.	Mar.
20	30	60

At the beginning of January there is a stock of 100 items and a closing stock of 140 items is required at the end of the quarter. Devise the production budget for the quarter so that a constant number of items is produced each month.

With these conditions, the budget might take the form:

	Jan.	Feb.	Mar.
Opening stock	100	130	150
Items produced	50	50	50
	150	180	200
Less sales	20	30	60
Closing stock	130	150	140

The opening stock of 100 items, together with a production of 50 items, enables the sales of 20 to be met for the month of January and give a closing stock of 130 items. This stock of 130 items, together with a production of 50 items, enables the sales of 30 items to be met in February and give a closing stock of 150 items. This stock of 150 items, together with a production of 50 items, enables the sales of 60 items to be met in March and give a closing stock of 140 items. A steady production rate of 50 items per month thus enables the sales to be met and the closing stock to be obtained at the end of the quarter.

6.2.3 Direct materials budget

On the basis of the production budget, a *direct materials budget* can be prepared. This indicates the direct materials that will be required to produce the items of the products listed in the production budget. The following budget illustrates this.

Direct materials budget for year

Material	Product X			Product Y			Product Z			Total units	Total unit price £	Total £
	Units	Unit price £	Total £	Units	Unit price £	Total £	Units	Unit price £	Total £			
A[1]	6 300	2	12 600	5 800	2	11 600	6 000	2	12 000	17 800	2	35 600
B[2]	12 600	1	12 600	2 900	1	2 900	6 000	1	6 000	21 500	1	21 500
			25 600			14 500			18 000			57 100

1. Product X requires for each item 1 unit of material A, product Y 2 units of material A and product Y 3 units of material A.
2. Product X requires for each item 2 units of material B product Y 1 unit of material A and product Y 3 units of material A.

In the above budget the number of units of material used for a product is obtained by multiplying the number of units of each material used per item produced, the number of items being obtained from the production budget for that period. The total estimated cost of a particular material for a particular product is obtained by multiplying the cost per unit of material by the number of units of material used for that product.

From the direct materials budget we can produce a *direct materials purchasing budget*. This will take account of stocks of materials existing at the beginning of the period and stocks which will exist at the end of the period.

Direct materials purchase budget for year

	Material A units	Material B units
Production requirements	17 800	21 500
Planned closing stock of materials	5 000	6 000
	22 800	27 500
Less planned opening stock	4 000	5 000
Total units to be purchased	18 800	22 500
Unit purchase price	£2	£1
Total purchases	37 600	22 500

The above budget enables the purchasing section to plan its budget for the period.

6.2.4 Direct labour budget

Direct materials are only one of the cost elements involved in production. Another element is direct labour. The following might thus be the *direct labour budget*.

Direct labour budget for year

	Product X	Product Y	Product Z	Total
Production (items)	6 300	2 900	2 000	
Hours per item	10	4	10	
Total budgeted hours	63 000	11 600	20 000	94 600
Wage rate per hour	4	4	4	
Total wages	252 000	46 400	80 000	378 400

In the above budget, the number of items to be produced is obtained from the production budget. The total number of hours budgeted for a product is obtained by multiplying the number of items of that product to be produced by the number of direct labour hours planned for it. The total wages for the direct labour employed on a product is obtained by multiplying the total number of direct labour hours planned for that product by the direct labour rate per hour for the labour used with that product.

6.2.5 Overheads budget

The three cost items needed to fully cost products are the direct materials costs, direct labour costs and overhead costs. The form of the budget for the *production overhead costs* will depend on the method used for the absorption of such costs. The following illustrates one approach.

Production (factory) overheads budget for year

	Overhead rate £/direct labour hour			Overheads £			Total
	Product X	Product Y	Product Z	Product X	Product Y	Product Z	
Variable overheads							
Indirect materials	0.10	0.20	0.10	6 300	2 320	2 000	
Indirect labour	0.20	0.20	0.20	12 600	2 320	4 000	
Power	0.10	0.20	0.10	6 300	2 320	2 000	
Maintenance	0.05	0.05	0.10	3 150	580	2 000	
				28 350	7540	10 000	45 890
Fixed overheads							
Depreciation				20 000	10 000	6 000	
Insurance				1 000	500	400	
Supervision				30 000	10 000	4 000	
				51 000	20 500	10 400	81 900
Total overhead				79 350	28 040	20 400	127 790
Rate/ labour hour				£12.60	£12.09	£10. 20	

The policy has been adopted in preparing the overhead budget of absorbing the overheads according to the number of direct labour hours budgeted for the product, these hours being obtained from the direct labour budget. The last row of the budget gives an overall budgeted production overhead rate that can be charged per hour of direct labour. It is obtained by dividing the total of the variable and fixed overheads by the number of direct labour hours budgeted for a product.

6.2.6 Cash budget

A particularly important budget for a business is the *cash budget*. This gives the plan necessary for the business to have sufficient cash to meet its commitments as they fall due. The cash budget considers only the actual flow of cash inwards and outwards from the business. The following illustrates the type of information contained in such a budget.

Cash budget for the year

	1st Quarter £	2nd Quarter £	3rd Quarter £	4th Quarter £	Total £
Opening balance	10 500	63 900	83 900	177 050	
Receipts from debtors	350 000	200 000	300 000	350 000	1 290 000
	360 500	263 900	383 900	527 050	
Payments:					
Purchase of materials	14 000	12 000	18 100	16 000	60 100
Payment of wages	94 600	85 000	95 000	103 800	378 400
Factory overheads	30 000	29 000	34 750	34 040	127 790
Selling and administration overheads	43 000	42 000	44 000	45 000	174 000
Capital expenditure				200 000	200 000
Dividends on shares	100 000			100 000	200 000
Others costs	15 000	12 000	15 000	18 000	60 000
	296 600	180 000	206 850	516 840	1 110 290
Closing balance	63 900	83 900	177 050	50 210	179 710

6.2.7 Master budgets

The above budgets enable control to be exercised in each of these spheres. The budgets, as indicated above, have obviously to be co-ordinated with each other, e.g. the production budget will obviously have to interlock with the sales budget, the stores budget with the production budget. When all the budgets for all functions have been prepared then a *master budget* can be prepared. This consolidates all the other budgets and will show the planned *profit and loss account* for the business as a whole and a *balance sheet*.

The following examples of profit and loss account and balance sheet illustrate the form such a master budget might take.

Profit and loss account for the year

	£	£	£
Sales			1 290 000
Costs of materials used		60 100	
Costs of direct labour		378 400	
Factory overheads		127 790	
Total manufacturing costs		566 290	
Add value of opening stock of finished goods	90 000		
Less closing stock of finished goods	120 000	(30 000)	
			536 290
Gross profit			753 710
Selling and administration costs			174 000
Budgeted operating profit for the period			579 710

The sales data, costs of materials used data, costs of direct labour data, factory overheads data used in the profit and loss account all come from the budgets given earlier in this chapter. The business starts with some stock of finished goods and plans to finish the period with some finished goods stock (see the production budget). These stocks have value and so must be taken into account. This value might be arrived at by using the standard cost, first-in-first-out, last-in-last-out, highest-in-first-out, average cost, market price or replacement price methods discussed in section 5.4. Since there is a greater value of closing stock than starting stock, the figure given in the second column is in brackets. This is because it is a negative cost value. Thus not all the manufacturing costs are expended on items that are sold. In the above example the selling and administration costs have been deducted from the gross profit, rather than being absorbed into the manufacturing costs.

The *balance sheet* is a statement of the financial position of a business at a given date. It gives the book value of all the assets, liabilities, share capital and reserves. The term current assets is used for those assets which are held for not longer than one year, fixed assets being those held for longer periods. Current liabilities are, like current assets, short-term. The balance sheet indicates where the business has obtained its capital and thus lists such items as the number and nominal value of shares held by the public and other organizations. Thus a balance sheet might appear as shown below.

Budgeted balance sheet for end of year

	£	£
Fixed assets:		
Land		320 000
Buildings and equipment	550 000	
Less depreciation	86 000	464 000
		784 000
Current assets:		
Materials stock	11 000	
Finished goods stock	120 000	
Debtors	40 000	
Cash	179 710	
	350 710	
Current liabilities:		
Trade creditors	65 000	285 710
		1 069 710
Financed by:		
Shareholders with 200 000 shares of £2 each	400 000	
Reserves	90 000	
Profit and loss account	579 710	1 069 710

6.3 Flexible budgets

The budgets so far considered can be termed *fixed budgets*. These assume that the volume of sales and the consequential activities in production can be estimated with a reasonable degree of accuracy. This might not be the case and there thus might be doubts as to what sales to plan for. In order that plans can still be made, *flexible budgets* are used in such circumstances. Flexible budgets give data for a number of different levels of activity. Thus flexible budgets might be used to give a sales budget and associated production budgets for a number of possible sales forecasts. Flexible budgets might also occur when costs are difficult to forecast. Thus a materials purchase budget might, if material prices are difficult to forecast, give data for a number of different prices for a material. Flexible budgets enable managers to compare actual costs with forecast costs under the conditions which they are actually working at. Thus, for example, a production department, e.g. a machining department, might have a flexible budget of the form:

Production department X flexible budget for some period

Item of expense	Budget formula	Activity level in number of output items		
		9 500	10 000	10 500
		£	£	£
Direct materials	£5.50 per output item	50 000	55 000	60 000
Direct labour	£10 per output item	90 000	100 000	110 000
Total direct costs		140 000	155 000	170 000
Overheads:				
Rent	Fixed overhead	5 000	5 000	5 000
Insurance	Fixed overhead	1 500	1 500	1 500
Lighting and heating	Fixed overhead	1 000	1 000	1 000
Power	Variable at £1 per output item	9 500	10 000	10 500
Maintenance	£500 fixed + £0.10 per output item	1 450	1 500	1 550
Indirect labour	£20 000 fixed + £1 per output item	19 500	30 000	30 500
Indirect materials	£0.10 per output item	950	1 000	1 050
Total overhead		38 900	50 000	51 100

6.4 Preparing budgets

There are a number of stages in the preparation of budgets. These can be considered to be:

1. Establishment of budget policy and guidelines for budget preparation. The budget policy will have to indicate the long-term plans of the management. Such plans might, for example, be that the product range is to be changed. The guidelines will have to indicate how allowances are to be made for possible price and wage increases and how such costs as those of the employer's contribution to pensions are to be dealt with.
2. Determination of the sales budget. This is the plan which affects and determines all the other budgets. The estimated volume of sales and the product mix determines what level of activity and what products the production department should aim to produce.
3. Initial preparation of production and other functional budgets.
4. Co-ordination and review of budgets.
5. Preparation of the master budget.
6. Consideration and acceptance of budgets by senior management.
7. Ongoing review of budgets.

6.4.1 Preparing a sales budget

The sales budget is a key budget. For the sales budget to be prepared, forecasts of sales over the budgetary period are required. Such forecasts are generally based on information from a variety of sources:

1 *Opinions of salespeople and sales managers* They might be asked to list for their sales area, for each product, last year's sales, the average over a number of years, the trend over the last few years and then indicate their views as to the sales for the next year.
2 *Market research* Data might be collected by a market survey in which potential customers are interviewed with regard to their future purchasing intentions. What do they want, at what price and in what quantities?
3 *General trade data* General trends in trade are likely to depend on the economic and political situation and thus a consideration of government publications and the financial press can provide useful data.

Example

In collecting data from which to prepare the sales budget for product Alpha, data of previous years sales were gathered and the opinions of the sales representatives obtained regarding their views for the coming year. The following is the data. Suggest a possible number of items that could be used in the budget for product alpha for the forthcoming year of 19X5.

19X1	19X2	19X3	19X4
1200	1500	1300	900

The views of the sales representatives were that the product was proving difficult to sell since it was considered to be out-of-date when compared with newer products that other companies had produced and they envisaged sales falling

The data indicates that the sales figure that should be considered for the year 19X5 should be less than 900. The question is as to how low the sales will drop. From the peak of 19X2 to 19X3 the sales dropped by 200 items, from 19X3 to 19X4 they dropped by 400 items. If this trend continues of the drop doubling each year then the 19X4 sales should drop by 800 to mean a forecast of just 100 items for 19X5. It might be considered that there will be a basic residual market for the product and this rate of decline will not continue. Thus a forecast might be for sales of perhaps 300 items.

6.5 Monitoring

In any control system there is an input to the system of a set value of the required outcome. The output of the system is then fed back to be compared with the input. This applies, for example, to a domestic central heating system controlled by a thermostat to give a set room temperature (figure 6.1). If there is a difference between the actual room temperature and that set by the thermostat then the error signal, i.e. the difference between the actual and set temperatures, sets into motion activity to remedy the error. Thus when the actual temperature is below the required temperature then the error signal switches on the boiler to pass hot water through the radiators.

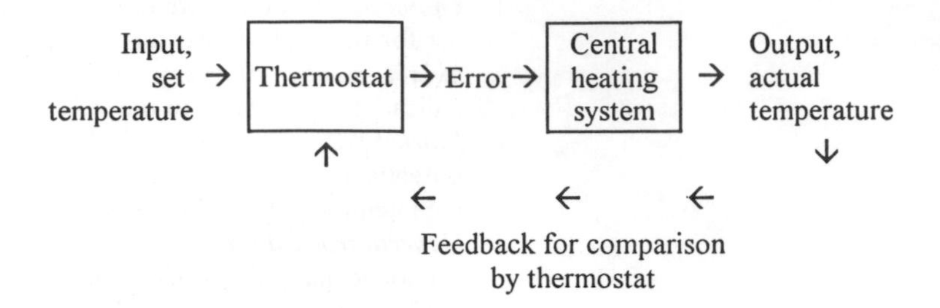

Figure 6.1 Central heating system

We have a similar system for the control of activities in an organization. The input is the budget, or budgets, and the output are the actual results. The output is then compared with the input. The difference between actual and budgeted performance is known as the *variance* (it is also used for the difference between actual and standard costs). The variance is the amount by which the costs vary.

Variance = actual performance – budgeted performance

A negative variance means that the amount indicated is that much under the budgeted amount, a positive variance that it is that much over the budgeted amount. Thus, for example, we might have for a production budget:

	Actual	Budget	Variance
Product X, units	10 300	12 000	– 1 700
Product Y, units	6 400	5 500	+ 900
Product Z, units	9 000	9 000	zero

Figure 6.2 illustrates the budgetary control process for an organization. Just as with the domestic central heating where the thermostat compares the actual temperature with the set temperature and initiates action if there is error, so with the organization the manager compares the actual activity with the budgeted activity and initiates action if there is variance.

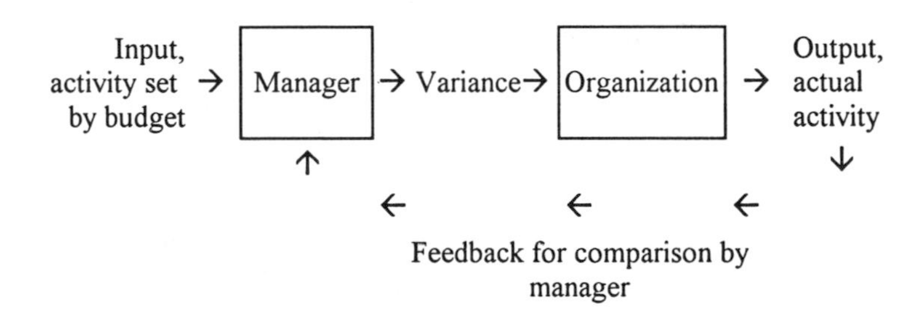

Figure 6.2 The budgetary control process

Example

The direct materials purchase budget for a material for the year was for 5000 units based on a standard costing of £10 per unit.
(a) What is the variance if the actual price per unit paid during the year was £9 per unit and the actual usage of the material was 5000 units?
(b) What is the variance if the actual price per unit paid during the year was £9 per unit and the actual usage of the material was 6000 units?

(a) The cost in the budget of the material is $5000 \times 10 = £50\ 000$. The actual cost of the material is $5000 \times 9 = £45\ 000$. Thus the variance is

$$\text{variance} = \text{actual performance} - \text{budgeted performance}$$

$$= 45\ 000 - 50\ 000 = -\ £5000$$

(b) The cost in the budget of the material is $5000 \times 10 = £50\ 000$. The actual cost of the material is $6000 \times 9 = £54\ 000$. Thus the variance is

$$\text{variance} = \text{actual performance} - \text{budgeted performance}$$

$$= 54\ 000 - 50\ 000 = £4000$$

Example

A product requires 2 units of material A, 3 units of material B and 1 unit of material C. The number of items forecast for the year is 200. The budget for the direct materials costs of the product for the year for 200 items gave:

Material A		Material B		Material C		Total
Units	Cost/ unit £	Units	Cost/ unit £	Units	Cost/ unit	£
400	5	600	2	200	10	5200

The actual number of items produced in the year was 220 and the actual costs of the materials per unit were material A £4, material B £3 and material C £9. Hence determine the variance in the direct materials cost for the product.

The actual data for the year is:

Material A		Material B		Material C		Total
Units	Cost/ unit £	Units	Cost/ unit £	Units	Cost/ unit	£
440	4	660	3	220	9	5720

The variance is thus

variance = actual performance – budgeted performance

= 5720 – 5200 = £520

6.6 Cost audits

The term *cost audit* is used for the verification of cost records and accounts and a check that the prescribed costing plan is being adhered to. The intention is to check that errors are not occurring or fraud. Such an audit might involve:

1. Checking the stocks of materials and goods in store and comparing the outcome with the figures that stock records indicate should be there.
2. Checking materials requisition forms and production orders and so checking that stock records have been correctly kept.
3. Checking that the prescribed procedures are used for charging materials withdrawn by production to production.
4. Checking the calculation of wages against clock cards and rates of pay.
5. Checking that labour hours allocated to jobs accurately reflect the true hours spent on the jobs.
6. Checking that the costs for indirect labour and indirect materials are legitimate.
7. Checking that overheads are apportioned to cost centres and products according to the prescribed procedures.

Such an internal audit should not be confused with the statutory audit which is carried out by external auditors. Under the Companies Act a company is obliged to employ independent auditors to audit the published accounts. There is, however, no obligation to have an internal cost audit.

Problems

Revision questions

1. What are the functions of budgets?

2. State the type of information that would be contained in:
   (a) a sales budget,
   (b) a direct materials budget,
   (c) a production overheads budget,
   (d) a master budget.

3. Explain what is meant by the terms fixed budget and flexible budget.

4. What is shown by a balance sheet?

5. Explain the term variance.

6. Explain what is meant by a cost audit.

Multiple choice questions
For problems 7 to 19, select from the answer options A, B, C or D the one correct answer.

7 Decide whether each of these statements is True (T) or False (F).

(i) Budgets are forecasts about what a business will achieve.
(ii) Budgets are plans about what a business will achieve.

Which option BEST describes the two statements?

A (i) T (ii) T
B (i) T (ii) F
C (i) F (ii) T
D (i) F (ii) F

Problems 8-10 relate to the following information:

A manufacturing company has a number of budgets. These include:

A Sales budget
B Direct materials budget
C Production overheads budget
D Profit and loss account

Select the budget which would BEST give the following information:

8 The total profit the company plans to earn in the year.

9 The materials the company plans to use in the manufacture of its products.

10 The total manufacturing costs.

Problems 11-12 relate to the following information:

The sales budget for a company includes the following information for one of its products:

Product Z
Units sold 4000
Selling price £100
Total revenue £400 000

11 The direct materials budget indicates that each item of product Z will require 3 units of material at a unit price of £2. The total cost of the direct materials for product Z will be:

A £6
B £8 000
C £12 000
D £24 000

12 The direct labour budget indicates that each item of product Z requires 10 hours of labour at a labour rate of £5 per hour. The total cost of the direct labour for the product will be:

A £50
B £20 000
C £40 000
D £200 000

13 Decide whether each of these statements is True (T) or False (F) F.

(i) The master budget summarises all the functional budgets.
(ii) The cash budget is to ensure that sufficient cash is available at all times.

Which option BEST describes the two statements?

A (i) T (ii) T
B (i) T (ii) F
C (i) F (ii) T
D (i) F (ii) F

14 As one of its budgets a business will prepare a profit and loss account.

Decide whether each of these statements is True (T) or False (F) F.

(i) The profit and loss account includes a list of the values of fixed assets of the business.
(ii) The profit and loss account includes a list of direct and indirect costs.

Which option BEST describes the two statements?

A (i) T (ii) T
B (i) T (ii) F
C (i) F (ii) T
D (i) F (ii) F

15 In the production budget of the ABC Engineering Company the budget for is for 8500 items of product X to be sold at a selling price of £100 per item. Product X uses two units of material P and one unit of material Q. Material P is costed at a rate of £5 per unit and material Q at £3 per unit. For the production department which makes product X, the direct materials usage budget is of the form:

	Units	Unit price	Total
Material P			
Material Q			
Total direct materials cost			x

The total that will appear at x is:

A £8
B £13
C £65 000
D £90 500

16 For the ABC Engineering Company, the sales budget for the year includes the entry:

Product X
Units sold 5000
Selling price £10
Total revenue £50 000

The production budget for the same period is of the form:

Product X	
Units to be sold	...
Planned closing stock	300
Total units required for sales and stocks	...
Less planned opening stock	200
Units to be produced	x

The total that will appear at x is

A 4500
B 4900
C 5300
D 5500

Problems 17-19 relate to the following information:

A business makes two products, X and Y. The budgeted sales and direct costs for the products are:

Budgeted sales in units, X: 3000, Y: 4000 units
Budgeted direct materials consumption per unit, X: 5 kg, Y: 2 kg
Budgeted direct materials cost: £3 per kg
Standard hours per unit of product, X: 4 hours, Y: 5 hours
Labour rate £5 per hour

17 The total budgeted direct costs of product X is:

A £ 45 000
B £ 60 000
C £ 63 000
D £105 000

18 The actual sales of the products are X: 2900 units, Y: 4100 units. The variance in the direct materials costs for the two products is thus:

A − £1000
B − £400
C + £400
D + £1000

19 A wage settlement results in the labour rate being increased by an average of 10% for the budget year. The total direct costs for product Y will thus increase, to the nearest percentage point, by:

A 5%
B 7%
C 8%
D 10 %

Assignments

20 This is a role playing assignment in which you and your fellow students are to play roles in a hypothetical company. Suppose you are all employees in a company making a single product. Select a product, but keep it simple to avoid over-complicating the assignment. Possible products might be a mains electric plug, a car component, or perhaps Christmas cards. One student, or a group, is to be responsible for devising the sales budget, another the production budget, another the direct materials budget, another the direct labour budget, another the production overheads budget, and another the master budget. Your budgets should be consistent with each other.

21 Examine government statistics for the last few years and, on that basis, make forecasts about the trends that might be occurring with regard to the purchasing of some type of domestic equipment, e.g. video recorders or washing machines.

22 Design a market research questionnaire that could be used to establish the type of market that might exist among students for sets of printed lecture notes.

23 Obtain a copy of a balance sheet for a company and explain the significance of all the items listed in it.

Case studies

24 The ABC Electronics Company assembles computers and sells them by mail-order. The sales manager has obtained the following data relating to the sales and distribution department.

Sales forecasts for the year:
- Product X: 140 000 units at £12 per unit
- Product Y: 132 000 units at £25 per unit
- Product Z: 36 000 units at £45 per unit

Sales costs forecasts for the year
- Salary of office manager £25 000
- Salaries of office staff £122 000
- Salaries of salespeople £100 000
- Sales promotion and advertising costs £40 000
- Sales office post, telephone, etc. expenses £12 000
- Sales office overheads £12 000
- Commission £40 000
- Salespeople expenses £30 000

(a) Present the information as a budget for that department.
(b) For the budget arising from the above data, some variance was found to occur by the end of the year. Produce a table showing the variances in volumes and values in the light of the following data:

- Sales of product X were down by 10%.
- Sales of product Y were up by 5%
- Sales of product Z were down by 2%
- Due to one salesperson leaving during the year and not being replaced, the salaries of the salespersons were down by £5000, all other salaries being as budgeted.
- Sales promotion and advertising were increased during the year to counteract the drop in sales of product X and so an increase of 10% over that budgeted occurred.
- Commission was down by 10%.
- Salespersons' expenses were down by 10%.
- All other data was as budgeted.

25 The production department of the ABC Engineering Company uses the following formulas when preparing its overhead costs budget:

- Direct labour rate £4 per hour
- Indirect labour £0.50 per direct labour hour
- Indirect materials £0.40 per direct labour hour
- Personnel and welfare services 5% of direct and indirect labour costs
- Semi-variable costs of electricity, maintenance, etc. 30% of direct labour hours
- Fixed costs of rent, insurance, etc. £50 000

(a) Prepare a flexible overhead costs budget for the department for levels of activity of 80%, 90% and 100% if 100% activity for the department involves 50 000 direct labour hours.

(b) Wage negotiations will take place during the period covered by the budget and the outcome is not clear. Prepare a flexible budget for 100% activity for wage settlements of increases of 2%, 5% and 7% in the direct labour rate.

26 The XYZ Electronics Company makes three products, X, Y and Z. The following is some of the information that was gathered for the preparation of budgets.

Sales forecasts
- Product X: 1000 units at £120 each
- Product Y: 2000 units at £140 each
- Product Z: 1400 units at £170 each

Direct materials
- Material M1 costs £4 per unit
- Material M2 costs £6 per unit
- Material M3 costs £10 per unit

- Product X uses 4 units of M1 and 2 units of M2
- Product Y uses 3 units of M1 and 4 units of M3
- Product Z uses 1 unit of M1, 2 units of M2 and 1 unit of M3

Direct labour
- Product X requires 10 hours per unit
- Product Y requires 15 hours per unit
- Product Z requires 20 hours per unit
- Labour is costed at the average rate of £5 per hour

Stocks existing at the beginning of the year
- Product X 1000 units
- Product Y 1400 units
- Product Z 400 units
- Material M1 20 000 units
- Material M2 14 000 units
- Material M3 5 000 units

Stocks forecast for end of the year
- Product X 900 units
- Product Y 1000 units
- Product Z 500 units
- Material M1 15 000 units
- Material M2 12 000 units
- Material M3 6 000 units

(a) Prepare fixed budgets for the year for sales, production quantities, direct materials usage, direct materials purchases, and direct labour.

(b) Prepare flexible budgets for the year for sales, production quantities, direct materials usage, direct materials purchases, and direct labour, assuming that the sales of unit figures could vary by plus or minus 10%.

27 The Alpha Engineering Company makes a single product. The marketing department have forecast that the sales for the next year will vary from month to month and be:

Jan: 10 000 units, Feb: 12 000 units, Mar: 15 000 units,
Apr: 12 000 units, May: 12 000 units, Jun: 12 000 units,
Jul: 10 000 units, Aug: 10 000 units, Sept: 11 000 units,
Oct: 12 000 units, Nov: 13 000 units, Dec: 9 000 units.

The product uses 3 units of material M1, 2 units of material M2 and 1 unit of material M3. Material M1 costs £4 per unit, material M2 £5 per unit and material M3 £10 per unit. The company adopts a policy of maintaining stocks of materials such that the closing stock at the end of a month is sufficient to meet the anticipated production requirements in the following month.

(a) Prepare a materials usage budget and a materials purchasing budget.
(b) At the end of the first quarter the actual sales were found to be:

Jan: 9500 units, Feb: 11 000 units, Mar: 14 000 units

Prepare a table showing the variances for the materials usage.

28 The ABC Plastics Company makes plastic sheeting in three grades, namely A, B and C. Prepare a materials usage budget and a materials purchasing budget on the basis of the following data:

Materials costs
M1: £0.30 per kg, M2: £0.40 per kg, M3: £0.35 per kg

Materials usage per roll of sheet
Grade A: 50 kg M1, 50 kg M2
Grade B: 60 kg M1, 40 kg M2, 30 kg M3
Grade C: 40 kg M1, 60 kg M3

Forecast sales for year
Grade A: 2000 rolls, Grade B: 1500 rolls, Grade C: 2500 rolls

Initial stocks
M1: 30 000 kg, M2: 40 000 kg, M3: 10 000 kg
Grade A: 200 rolls, Grade B: 120 rolls, Grade C: 200 rolls

Forecast stocks at end of year
M1: 30 000 kg, M2: 20 000 kg, M3: 5 000 kg
Grade A: 120 rolls, Grade B: 100 rolls, Grade C: 140 rolls

29 The Car Components Company makes two products, namely X and Y. Both the components are made from the same basic raw material, using the same labour force. The budgeted sales and costs for the two components for the forthcoming year are:

Sales
    Product X: 48 000 items at £12 per unit
    Product Y: 40 000 item at £20 per unit

Production
    Materials, X: 3 units per item, Y: 2 units per item
    Labour, X: 2 hours, Y: 4 hours

Costs
    Direct materials £2 per unit
    Direct labour £5 per hour
    Variable overheads are costed at £1 per labour hour
    Fixed overheads are £10 000

Initial stocks
    X: 4000 items, Y: 3000 items, material 10 000 units
    Stocks planned for end of budget year
    X: 5000 items, Y: 4000 items, materials 12 000 units

(a) Prepare the direct labour and the materials purchase budgets.
(b) A productivity agreement results in the direct labour rate increasing by £0.40 per hour and the standard times for each product being reduced by 20%. What will be the net cost/saving to the company of this agreement?

7 Inventory control

7.1 Inventory control

Consider a simple situation of a small business and the type of inventory problems that might occur. Suppose the business is a stall on a market which sells cups of coffee. The concerns might then be:

1 There must be enough coffee and plastic cups available to not hamper the production of cups of coffee. Thus there should be adequate stocks of coffee and cups to perhaps ensure that they will not run out of such materials on a market day.
2 They must not carry too much stock of coffee and cups. They have to pay for such stocks and a place to store them. There is no point in having a stock which is large enough to last a few months if the stocks can easily be replenished at a cost lower than the cost of the capital that would otherwise be tied up in stocks and involved in paying storage costs.
3 They brew up a pot or pots of coffee from which they pour the cups for customers. Customers expect to be served with a cup without any significant delay. This means that the business must carry a stock of "finished goods" from which customer demand can be met. How big a stock of finished goods, i.e. brewed coffee, should the business carry at any particular time? If they carry too much then it deteriorates and has to be poured away. If they carry too little then they cannot meet the customers' requirements for coffee without any significant delay. Too much brewed coffee costs money, because of the waste, and too little costs money in terms of lost orders.

The type of inventory concerns outlined in the above example are the type of inventory problems that have to be faced by any manufacturing organization, whether small or large.

Inventory control is concerned with the determination of appropriate inventory levels and the maintenance of these levels. Thus, the production department will be concerned that production is not hampered by a lack of required materials or components. When production needs, say, a particular quantity of a particular size of mild steel bar, then that material needs to be available. If it is not available at the required time then production might grind to a halt. There would also be concern that there was not a surplus of materials and components beyond that required by production, since the holding of stocks incurs costs as well as the money tied up in the stock. A business which sells items from stock will also be concerned with inventory levels of finished goods so that customers do not have to wait too long when they order goods. Finished goods stocks present problems. If the stock is too low then the time delay for customers might be too long, if the stock level is too high then too much cash is tied up in the stock and holding costs. The aim of inventory control is to obtain the necessary inventories at the minimum cost to the organization.

7.1.1 Inventory costs

Inventories incur costs. Consider the situation with respect to the stocks of raw materials, say mild steel bar. The costs can be considered to be the cost of the item itself, the costs associated with the replenishment of stocks (e.g. the cost of paperwork, transportation, etc.), and holding costs. The holding costs are the costs associated with keeping the items in store for a period of time and include the costs associated with the cost of the storage facilities, insurance costs, costs associated with the stock become obsolete and/or deteriorating, and the costs of the capital tied up in the inventory (e.g. the interest the money would have earned if left in the bank instead of having been used to buy the stock. A manufacturing business might have a large amount of capital tied up in inventories. The purpose of inventory control is to keep the costs of inventories down to the minimum that is commensurate with keeping production flowing without interruption and meeting the requirements of customers.

7.1.2 Inventory levels

To illustrate the problems associated with the control of inventories, consider a simple example of a business which starts off with a stock of, say, 100 units of some material. If there is a steady demand from production for that material then the stock of the material will steadily drop with time. Suppose the demand is forecast as being 20 units per week. Then we would have:

Week	1	2	3	4	5	6	7	8
Demand (units)	20	20	20	20	20	20	20	20
Stock at start of week (units)	100	80	60	40	20	0	(20)	(40)
Stock at end of week (units	80	60	40	20	0	(20)	(40)	(60)

After week 1 the stock will have dropped to 80 units, after weeks 2 to 60 units, after week 3 to 40 units, after week 4 to 20 units, after week 5 to 0, and thereafter there will be shortage of stock. Note that the convention is adopted in accounts of writing negative quantities in brackets, e.g. (20) is – 20. Thus in week 6 there is a shortage of stock of 20 units, in week 7 a shortage of 40 units. The above table indicates that the stock will run out at the end of week 5. Thus more stock will have to be ordered at some time. The business might adopt a policy of endeavouring to not let the stock drop below 20 units. Thus they might order stock of 80 units to arrive in week 4. If it takes a week from placing an order to it arriving, then the order would need to be placed in week 3. Then the stock levels become:

Week	1	2	3	4	5	6	7	8
Demand (units)	20	20	20	20	20	20	20	20
Stock at start of week (units)	100	80	60	40	100	80	60	40
Stock at end of week (units)	80	60	40	20 +80	80	60	40	20 +80
Stock received (units)				80				80
Stock ordered (units)			80				80	

Figure 7.1 shows graphically how the stock level might thus change with time. When the stock level reaches 20 units the stock is replenished back up to the 100 level by the arrival of an order of 80 units. The level of 20 units is referred to as a *buffer stock*, since it is a buffer against forecasts being wrong and stock being used at a faster rate than it planned. The larger the buffer stock the more cautious a business is regarding its forecasts and the more concerned it is about running out of stock. The business considers the costs of running out of stock to be greater than the holding costs involved in keeping a buffer stock. To obtain a delivery at the fourth week, an order has to be placed 1 week earlier. This 1 week is termed the *lead time*. Thus the stock is ordered when the stock level has dropped to 40 units. This method of ordering stock when a particular stock level is reached is termed *stock point generation*.

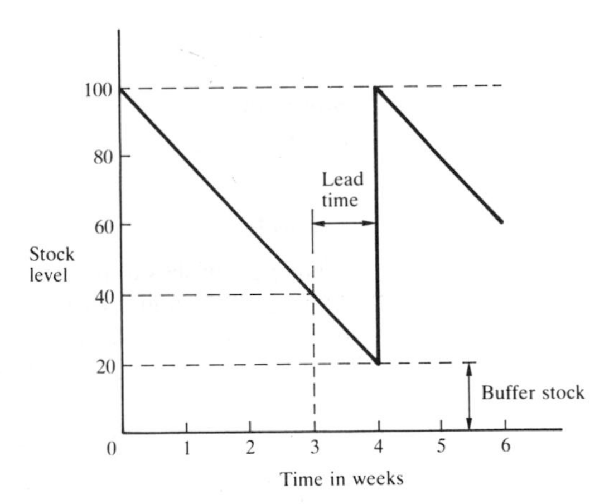

Figure 7.1 Stock levels

Example

At the beginning of the year a business has an opening stock of 120 units of finished goods. There is a steady demand for the product of 40 units per month. Production needs a month to produce a batch of 120 units. The business has a policy of not letting the stock of finished goods drop below 40 units. When should production be notified of the need to produce a batch?

With no production orders given the stock levels will be:

	Jan	Feb	Mar	Apr	May	Jun	July	Aug
Demand (units)	40	40	40	40	40	40	40	40
Stock at start of week (units)	120	80	40	0	(40)	(80)	(120)	(160)
Stock at end of week (units)	80	40	0	20	(80)	(120)	(160)	(200)

For the stock level not to drop below 40 units then there will have to be a delivery of finished goods by the end of February. Thus production will need to produce the goods in January. With batches of 120 units:

	Jan	Feb	Mar	Apr	May	Jun	July	Aug
Demand (units)	40	40	40	40	40	40	40	40
Stock at start of month (units)	120	80	160	120	80	160	120	80
Stock at end of month (units)	80	40 +120	120	80	40 +120	120	80	40 +120
Stock received (units)		120			120			120
Stock ordered (units)	120			120			120	

Example

A business wishes to have an even production flow throughout the year. At what rate should they produce finished units if at the beginning of the year there is an opening stock of 110 units, stock levels must not drop below 100 units, and the demand for units is forecast as being:

	Jan	Feb	Mar	Apr	May	Jun	July	Aug
Demand (units)	20	30	60	80	70	60	40	40

Since we start off with 110 units of stock and the forecast demand is 20 units in January, then to maintain the minimum stock level there must be a production of at least 10 units in that month. This would mean a stock level at the end of January of 100 units. But to maintain this minimum stock for the end of February the production would have to be 30 units. Suppose we consider a steady production of 30 units per month. Then:

	Jan	Feb	Mar	Apr	May	Jun	July	Aug
Demand (units)	20	30	40	60	70	60	40	40
Stock at start of month (units)	110	120	120	110	80	40	10	0
Stock at end of month (units)	90 +30	90 +30	80 +30	50 +30	10 +30	(20) +30	(30) +30	0 +30
Stock received (units)	30	30	30	30	30	30	30	30

With a production of 30 units per month then at the end of January there would be a stock of 120 units, at the end of February 120 units and at the end of March 110 units. But only producing 30 units in April will mean that the stock level at the end of that month will have dropped to 80 units. Thus the production level has to be set higher. Suppose production is set at 50 units per month. Then we have:

	Jan	Feb	Mar	Apr	May	Jun	July	Aug
Demand (units)	20	30	40	60	70	60	40	40
Stock at start of month (units)	110	140	160	170	160	140	130	140
Stock at end of month (units)	90 +50	110 +50	120 +50	110 +50	90 +50	80 +50	90 +50	100 +50
Stock received (units)	50	50	50	50	50	50	50	50

The result meets all the demands and gives a stock level which is maintained above the minimum stock level specified. However, the stock levels are generally significantly above the minimum 100 units level and we could probably find a lower rate of production which would meet the requirements. With 45 units per month we have:

	Jan	Feb	Mar	Apr	May	Jun	July	Aug
Demand (units)	20	30	40	60	70	60	40	40
Stock at start of month (units)	110	135	150	155	140	115	100	105
Stock at end of month (units)	90 +45	105 +45	110 +45	95 +45	70 +45	55 +45	60 +45	60 +45
Stock received (units)	45	45	45	45	45	45	45	45

7.2 Economic order quantities

Consider a situation where a production department used 100 units of a material per week. Should the purchasing department order the material in lots of 100 a time and so have just enough stock in each order to cover a week or in lots of 1000 and have stock in each lot ordered to cover 10 weeks of use? How big an order should be placed so that the cost to the business is a minimum. The issue is one of balancing the costs associated with holding stock against those involved in purchasing it.

The factors in favour of placing large orders infrequently are:

1 Discounts may be available for larger quantities.
2 It may be easier to obtain large quantities rather than small quantities.
3 There is less chance of production running out of materials.

The factors in favour of small orders placed more frequently are:

1 Less cash tied up in stock in store.
2 Less storage space required.
3 Lower insurance costs.
4 Less chance of stock deteriorating or becoming obsolete.

We can develop a simple mathematical model to aid in determining how big an order to place and how frequently to order. Consider an inventory system for raw materials and components where a quantity Q of the item is ordered each time an order is placed. For simplicity we will assume:

1 We will assume that stock levels are allowed to fall to zero, i.e. there are no buffer stocks.
2 Shortages are not allowed, i.e. stock must be replenished at the precise moment when stock level reaches zero.
3 That the order is placed at such a time that deliveries are received at the precise moment the stock level reaches zero, all the order being delivered to the store at the same time.
4 The rate of withdrawal of items from the stock is constant. For example, the withdrawal of stock might be 100 units per day and assumed to remain at this level every day.
5 There is a fixed ordering cost for each lot ordered which is independent of the number of items in the lot. Thus there might be a cost of, say, £50 to make an order, whether it be for 1 unit of material or 100 units.
6 There is a fixed holding cost per item so that the cost of holding 10 items in stock is ten times the cost of holding 1 item in stock. Thus, for example, there might be a holding cost of £10 per item per year.

If the items are taken from the store, which was initially Q, at a constant rate, the way the stock, i.e. inventory, in the store varies with time will be of the form shown in figure 7.2. The stock level in the store will fall at a constant rate from Q to zero. Then an order will be received which will push the stock level back up to Q. It will then again fall at a steady rate to zero, when it again will be pushed back up to Q by receipt of another order, and so on.

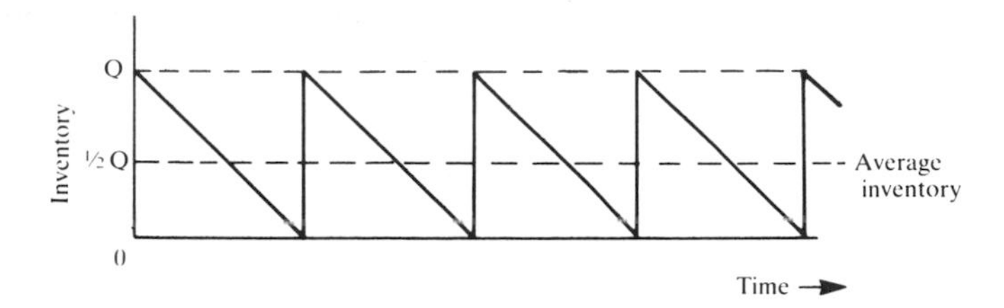

Figure 7.2 Variation of inventory level with time

The average amount of stock will be $\frac{1}{2}Q$. If C_h is the holding cost per item of stock per unit of time then the holding cost for the stock is $\frac{1}{2}QC_h$. For example, if the quantity held in stock falls at a steady rate from 1000 to 0 and the holding cost per item is £2, then the holding cost is

holding cost = $\frac{1}{2}QC_h = \frac{1}{2} \times 1000 \times 2 =$ £1000.

We can consider the holding costs in a different way. If c is the cost of each item of stock when bought then, assuming this cost is constant and there are no discounts for quantity, the average amount of money tied up in the inventory is $\frac{1}{2}Qc$. Thus, for example, if the ordering quantity Q is 1000 items and each costs £10 then the average cost is £5000. If C is the cost per unit of capital per unit time of carrying the items in stock, e.g. the rate of interest that could have been gained by the capital if it has not been used to purchase the stock, then the holding cost is $\frac{1}{2}QcC$. For example, if the interest rate is 0.5% per week then the lost interest on an average capital tied up in stocks of £5000 is £25. Thus $C_h = cC$.

If the stock is drawn from store at a constant rate r, the time taken for the stock to reach zero is Q/r. For example, if stock is withdrawn at the rate of 100 items per week then for an order size of Q = 1000, the time taken for the stock to reach zero will be 1000/100 = 10 weeks. If the cost of placing an order is C_s, then the ordering cost will be incurred each time the stock reaches zero. Thus the average ordering cost per unit time is rC_s/Q. For the example we have been using in this paragraph, if the cost of ordering an item is £50 then the average ordering cost per week is 50/10 = £5. The total cost of the inventory is the sum of the holding cost and the ordering cost:

inventory cost per unit time = holding cost + ordering cost

Thus

$$\text{inventory cost per unit time} = \tfrac{1}{2}QC_h + \frac{rC_s}{Q}$$

The holding cost increases in proportion to the quantity ordered. The ordering cost is inversely proportional to the quantity ordered. Thus graphs of these costs against the quantity ordered are of the form shown in figure 7.3. The total cost can be obtained by adding the holding cost and ordering cost graphs.

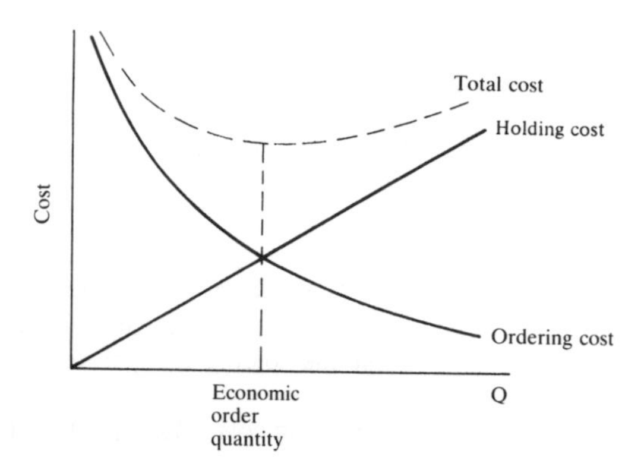

Figure 7.3 The costs

The total cost is a minimum when the holding cost equals the ordering cost (you can check this by making measurements on the graph or differentiating the inventory cost equation and so finding the condition for the cost to be a minimum), i.e. when

$$\tfrac{1}{2}QC_h = \frac{rC_s}{Q}$$

Hence the optimum quantity to order each time if the costs are to be kept to a minimum is given by

$$Q^2 = \frac{2rC_s}{C_h}$$

This quantity is known as the *economic order quantity* (EOQ). Thus

$$\text{EOQ} = \sqrt{\frac{2rC_s}{C_h}}$$

Alternatively we can write

$$\text{EOQ} = \sqrt{\frac{2rC_s}{cC}}$$

The above derivation assumed a constant rate at which items were drawn from stock. When the withdrawal rate r is not constant then the equation is still used but r is put equal to the average rate. The EOQ equation then gives a rough idea of the best order quantity.

Example

For a company, the cost per month of holding stocks of mild steel bar used in its production process is £1 for each unit of mild steel bar. Mild steel bar is drawn from store by the production department at the rate of 100 units per month. The cost of placing an order for mild steel bar is £10. What is the economic order quantity?

Using the symbols given above for the derivation of the EOQ equation, then we have C_h = £1 per unit per month, C_s = £10 and r = 100 per month. Thus

$$\text{EOQ} = \sqrt{\frac{2rC_s}{C_h}} = \sqrt{\frac{2 \times 100 \times 10}{1}} = 44.7$$

This figure amounts to just about enough stock to cover two weeks of production. If the cost of ordering had been higher, say, £50 then the EOQ would have become 100 and so enough stock would have been ordered to cover a month. The higher the cost of ordering the greater the amount of stock ordered at a time, the amount of stock ordered being inversely proportional to the square root of the holding cost.

Example

The production department uses 360 of a special type of integrated circuit per year. The cost of placing an order is £10. The cost of one of these integrated circuits is £8. If the interest rate that could have been earned by the capital is 25% per year, what will be the economic order quantity?

Using the symbols given above for the derivation of the EOQ equation, then we have C_s = £10, C = £8, c = 25% and r = 360 per year. Thus

$$\text{EOQ} = \sqrt{\frac{2rC_s}{cC}} = \sqrt{\frac{2 \times 360 \times 10}{0.25 \times 8}} = 60$$

Thus 60 items are ordered at a time. This means 360/60 = 6 orders per year.

7.2.1 Order times

Consider how the inventory level varies with time when there is a constant rate r at which items are drawn from stock, as illustrated in figure 7.4 (this is just a piece of the graph in figure 7.2).

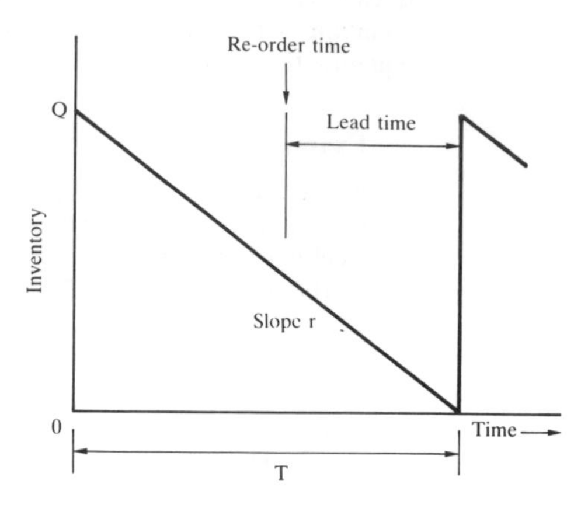

Figure 7.4 Time between orders

The time T taken for the stock to fall from Q to 0 is the time between which orders are delivered. Thus

$$\frac{Q}{T} = r$$

But Q is the economic order quantity (EOQ) and so is given by the equation developed in section 7.2. Hence

$$T = \frac{\text{EOQ}}{r} = \frac{1}{r}\sqrt{\frac{2C_s r}{C_h}} = \sqrt{\frac{2C_s}{rC_h}}$$

T is thus the time between orders of the economic order quantity. Thus to meet the demand, an order must be placed at periodic intervals through the year, the time between orders being T.

There will be some time needed between placing an order and the order being delivered. Such a time is referred to as the *lead time*. Thus if, for example, we have the time between orders of the EOQ as 10 weeks then with a lead time of 4 weeks the order would have to be placed 6 weeks after each delivery.

Example

For a company, the cost per month of holding stocks of mild steel bar used in its production process is £1 for each unit of mild steel bar. Mild steel bar is drawn from store by the production department at the rate of 100 units per month. The cost of placing an order is £10.

(a) What are the time intervals between orders for the economic order quantity?
(b) How soon after the receipt of an order will the next batch have to be ordered if there is a lead time of 0.2 of a month?

(a) This data is the same as used in the previous example. We thus have C_h = £1 per month, C_s = £10 and r = 100 per month. Hence

$$T = \sqrt{\frac{2C_s}{rC_h}} = \sqrt{\frac{2 \times 10}{100 \times 1}} = 0.45 \text{ months}$$

(b) The order must be placed 0.2 of a month before the order is to be received. Thus it must be placed 0.45 – 0.2 = 0.25 of a month after receipt of the last order.

7.3 Finished goods stocks

Consider the situation where a business makes a product for stock by batch production. Suppose, for example, the demand for a product is 1000 units per year and this production can be obtained by 2 months work. Should production produce all the 1000 units in the first two months of the year or, perhaps, produce the 1000 items in two batches with 500 in the first month and 500 in the sixth month? What determines the size of a batch?

The problem is one of inventory management. Production are delivering items to stock and then items are being drawn from stock against the orders of customers. Consider the situation where:

1 The stock level is allowed to drop to zero.
2 Shortages are not allowed, i.e. stock must be replenished at the precise moment when stock level reaches zero.
3 Items are delivered into stock as a complete batch, the deliveries being received at the precise moment the stock level reaches zero.
4 The rate at which items are withdrawn from the stock is constant. For example, the stock might be withdrawn at the rate of 100 units per day and assumed to remain at this level every day.
5 There is a fixed setting up cost for each batch made which is independent of the number of items in the lot. Thus there might be a cost of, say, £5000 to set up for production, whether it be for 1 unit of material or 100 units.
6 There is a fixed holding cost per item so that the cost of holding 10 items in stock is ten times the cost of holding 1 item in stock. For example, there may be a holding cost of, say, £10 per item per year.

Thus if we have a batch size of Q then the way the stock of finished items will vary with time will be of the form shown in figure 7.5. Note that this is the same as figure 7.2, the problem being essentially the same as that discussed in section 7.2. We can thus treat is in exactly the same way.

The average amount of stock will be $\frac{1}{2}Q$. If C_h is the holding cost per item of stock per unit of time then the holding cost for the stock is $\frac{1}{2}QC_h$. If the stock is drawn from store at a constant rate r, the time taken for the stock to reach zero is Q/r. If the cost of setting up for a batch is C_s, then the setting up cost will be incurred each time the stock reaches zero. Thus the average ordering cost per unit time is rC_s/Q. The total cost of the inventory is the sum of the holding cost and the setting up cost:

inventory cost per unit time = holding cost + setting up cost

Thus

$$\text{inventory cost per unit time} = \tfrac{1}{2}QC_h + \frac{rC_s}{Q}$$

The holding cost increases in proportion to the quantity ordered. The setting up cost is inversely proportional to the quantity ordered. Thus graphs of these costs against the quantity ordered are of the form shown in figure 7.5. Note that this graph is the same form as that given in figure 7.3. The total cost is obtained by adding the holding cost and setting up cost.

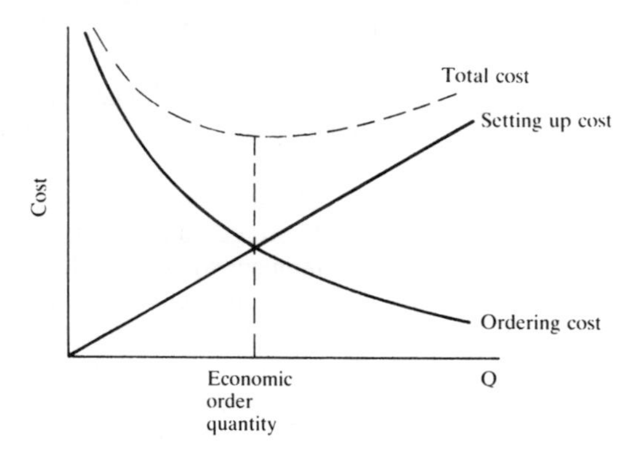

Figure 7.5 The costs

The total cost is a minimum when the holding cost equals the setting up cost (you can check this graphically or by differentiating the above equation to find when the gradient of the graph is zero), i.e. when

$$\tfrac{1}{2}QC_h = \frac{rC_s}{Q}$$

Hence the economic batch quantity EBQ is given by

$$Q^2 = \frac{2rC_s}{C_h}$$

and so

$$\mathrm{EBQ} = \sqrt{\frac{2rC_s}{C_h}}$$

This equation is the same form as that for the economic order quantity (see section 7.2).

Example

A production department has to supply 10 000 items of a particular component per year. Each production run has a machine setting up cost of £200. The cost of holding the stock is £4 per item per year. What is the batch size which will give the minimum cost?

Using the symbols given above for the derivation of the EBQ equation, then we have C_h = £4 per item per year, C_s = £200 and r = 10 000 per year. Assuming that the batches are delivered in their entirety into stock then

$$\mathrm{EBQ} = \sqrt{\frac{2rC_s}{C_h}} = \sqrt{\frac{2 \times 10\,000 \times 200}{4}} = 1000$$

Thus it is most economic to make the 10 000 items in batches of 1000.

7.4 Just-in-time

Strict control over inventory levels is necessary if money is not to be wasted by being tied up in holding costs, e.g. the costs of storage and of the interest lost on the capital sums used to pay for the stocks. An approach, however, that is frequently used by many companies is, because they are not sure of the demand, *just-in-case* buying and building up of stocks. They wish to make sure that whatever the demand, they will not run out of stock and so the machines and operators in production will be kept fully working. The cost of running out of materials is judged to be greater than the cost of holding materials in store.

A new approach which is becoming widely adopted is *just-in-time*. This involves only obtaining stocks when they are required. It means taking a risk that the demand will not have been accurately forecast and there might be insufficient stock to keep production working. This is a policy which originated in Japan. It has been suggested that the policy arose because of the shortage of space in Japan and hence the high cost of storage space.

Just-in-time is based on such requirements as:

1 *Good forecasting of demand* In order that materials can be delivered at just the right time, or customers receive their orders from a company at the right time, considerable attention must be devoted to forecasting demand.
2 *Stringent quality assurance* The materials received must, because they are delivered just at the moment they are required, be of the required

quality. Rejecting materials because they are not up to the required quality would present severe problems. Likewise, the products of a company, because they are only made so that they are ready at the instant when the demand for them occurs, must be of the required quality. Rejecting products at that stage would mean hold ups in the delivery of products to customers.

3 *Delivery dates must be met* Suppliers must meet delivery dates for materials. Likewise, the production department of a company must be able to deliver finished goods on time. In-house schedules must be maintained.

4 *Suppliers must be fully integrated within the operation* Because the suppliers of materials must conform to stringent quality specifications and meet strict delivery dates, it is essential that key suppliers are effectively integrated within the operation. There must be very good communications between the company and the suppliers. Computer-to-computer communications might well be used.

5 *Production must be economic with small batches* Small batches are likely to be the means by which many companies are able to deliver finished goods just-in-time for customers' orders. To make small batch production economic means utilizing flexible machines and using equipment to ensure that setting-up costs are kept to the absolute minimum.

Problems

Revision questions

1 Explain how inventories incur costs.

2 Explain what is meant by the term economic order quantity.

3 Calculate the economic order quantity if the cost of ordering is £20 per lot, the purchase price is £3 per item, the holding cost per year is 35% of the value of the stock and the rate of usage is 800 items per year.

4 Calculate the economic order quantity if the cost of ordering is £20 per lot, the holding cost is £10 per item per year and the rate of usage is 1000 items per year.

5 For a particular product the economic order quantity is 400 and the rate of usage is 100 per week. If the lead time for ordering is 1 week, how long after the receipt of an order should the next order be placed?

6 A production department has to supply 1000 items of a particular component per year. Each production run has a machine setting up cost of £240. The cost of holding the stock is £12 per item per year. What is the batch size which will give the minimum cost?

Multiple choice questions
For problems 7 to 18, select from the answer options A, B, C or D the one correct answer.

7 Raw materials and components needed for processing must be available to the production department at the right time and in the required quantities if production is to continue unhampered and production schedules met.

Decide whether each of these statements is True (T) or False (F).

(i) The quantity of raw materials and components held in store for production must be large enough to ensure a smooth production flow.
(ii) The quantity of raw materials and components held in store for production must be such that the capital tied up in the storage and holding costs is not greater than the costs incurred in running out of them and reordering them.

Which option BEST describes the two statements?

A (i) T (ii) T
B (i) T (ii) F
C (i) F (ii) T
D (i) F (ii) F

8 Businesses which meet customers' orders from stock tend to carry stocks of finished goods.

Decide whether each of these statements is True (T) or False (F).

(i) Stocks of finished goods should be kept as low as possible in order to keep costs down
(ii) Stocks of finished goods should be kept at high as is necessary to meet customers' requirements for prompt deliveries.

Which option BEST describes the two statements?

A (i) T (ii) T
B (i) T (ii) F
C (i) F (ii) T
D (i) F (ii) F

9 Decide whether each of these statements is True (T) or False (F).

Inventory control in a business is concerned with the control of the amounts of:
(i) the incoming raw materials and components,
(ii) part processed components between stages in production,
(iii) finished goods.

Which option BEST describes the above statements?

A (i) T, (ii) and (iii) F
B (i) and (iii) T, (ii) F
C (i), (ii) and (iii) T
D (iii) T, (ii) and (iii) F

10 Decide whether each of these statements is True (T) or False (F).

The cost of an inventory of raw materials for use in the production process is proportional to:
(i) the reorder cost per order placed
(ii) the holding cost per item
(iii) the rate at which the materials are drawn from stock.

Which option BEST describes the above statements?

A (i) F, (ii) F, (iii) F
B (i) F, (iii) T, (ii) F
C (i) T, (ii) T, (iii) F
D (iii) T, (ii) T, (iii) T

11 The economic order quantity is the order size which:

A Minimizes the holding cost
B Minimizes the ordering cost per order placed
C Minimizes the size of order to be placed
D Minimizes the inventory cost

12 The economic order quantity is doubled if

A The ordering cost is doubled
B The rate at which stocks are used is quadrupled
C The holding cost is quadrupled
D The purchasing price is doubled

13 The economic order quantity for the purchase of packets of particular resistors used in electronics assembly in a company is 1200. The rate of usage by the assembly production line is constant at a rate of 400 packets of these resistors per week. The lead time for the ordering is 1 week.

Decide whether each of these statements is True (T) or False (F).

(i) An order should be placed every 3 weeks.
(ii) An order should be placed 3 weeks after an order is received.

Which option BEST describes the two statements?

A (i) T (ii) T
B (i) T (ii) F

C (i) F (ii) T
D (i) F (ii) F

14 Figure 7.6 shows how, for a business, the stock level of a particular item varies with time.

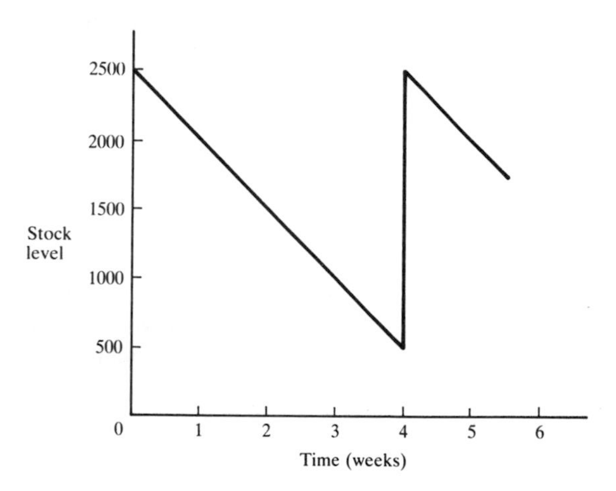

Figure 7.6 Problem 14

Decide whether each of these statements is True (T) or False (F).

(i) With a lead time of 1 week, stock is ordered when the stock level reaches 2000 items.
(ii) The quantity of stock ordered and delivered in week 4 is 2000 items.

Which option BEST describes the two statements?

A (i) T (ii) T
B (i) T (ii) F
C (i) F (ii) T
D (i) F (ii) F

15 Decide whether each of these statements is True (T) or False (F).

(i) Just-in-time means only starting to make a product when an order is received.
(ii) Just-in-time means forecasting when a product is required and making the product in anticipation of that demand.

Which option BEST describes the two statements?

A (i) T (ii) T
B (i) T (ii) F
C (i) F (ii) T
D (i) F (ii) F

16 Reducing the setting-up time for machines used to produce a particular product directly results in:

A A reduction in economic batch quantity
B A reduction in the holding cost
C A reduction in the total cost of an item
D A reduction in the rate at which items are produced

17 For the first three months of the year the forecast sales of a product are: January 50 units, February 60 units, March 70 units. There is an opening stock of 80 units. If stock levels are not to drop below 50 units, which of the following rates of production per month will ensure that the demand is met with the minimum of capital tied up in the stock?

A 30 units
B 50 units
C 60 units
D 70 units

18 The demand for steel bar by the production department is constant at the rate of 20 units per month. If at the beginning of January there is a stock of 100 units and orders have a lead time of one month, by the end of which month should an order be placed for 80 units if the stock level is not to drop below 20 units?

A January
B February
C March
D April

Assignments

19 Many colleges have shops which sell stationery to students. Investigate for such a shop how the stock of a particular type of stationery varies with time. Obtain the appropriate data to make an estimate of the economic order quantity. Compare your result with the orders actually placed by the shop.

20 This a simple role-playing exercise. Consider the production of a simple Christmas card based on a single sheet of paper which has to be folded into two, a Christmas tree in green with a few ornaments in other colours drawn on the front, the words "Happy Christmas" written on

the front in red, and finally the words "A Happy Christmas and a prosperous New Year" written in black on the inside. The sheets of paper have to be drawn from stock. Set up a production line involving four students for the four stages of operation, providing the necessary coloured pens, and another student in charge of the paper stocks.

(a) Run the production for, say, 10 minutes with the stores having initially an unlimited supply of paper in stock. Time the various stages of the production and determine the rate at which paper is withdrawn from stock. Discuss the operation. Does production run smoothly with no inventories of part-finished goods or hold ups occurring between production stages? If such occur, are there ways by which you can eliminate them?
(b) Consider now the situation where stores have to order the paper. If stores have not the space to carry more than a stock of 10 sheets of paper, how often would they need to reorder?
(c) Consider how the stores and production operations might change if a just-in-time policy was adopted? Consider this for the paper stocks, the part-finished inventory between production stages and the finished goods inventory.

Case studies

21 The management of the ABC Engineering Company is concerned with the increases in costs that have occurred since the production department switched to a production line for the assembly of its products. The production line consists of a line of workers with each worker carrying out a specific job on the product. Because the complexity of the tasks carried out, and the effectiveness of the workers in the line, are not all the same, buffer inventories of part-finished products have been established between each worker in the line so that all the workers are fully employed all the time. Management feels that this has resulted in large amounts of capital becoming tied up in the buffer stocks.

(a) Explain why management feel that the increase in costs might be attributed to the buffer stocks.
(b) How would you suggest that the costs be reduced?

22 The XYZ Engineering Company buys screws from an outside supplier. The production department feels that the minimum stock level of a particular screw that should be held is 100 000 screws, while the maximum stock at any instant should be 700 000 screws. The screws are used at a steady rate of 120 000 per month. The lead time from placing an order to receiving the screws is two months.

(a) Plot a graph showing how the stock of screws will vary with time.
(b) What is likely to be the level of screws in store when reordering has to occur?

(c) Discuss the policy of the company with regard to its inventory of screws and consider whether there could be a more economic policy (for data you might need, think of possible figures). The production department consider that any other policy would be a waste of effort considering the cheapness of the item.

23 The production manager of the ABC Engineering Company is trying to work out the production level required for product A during the first six months of the year in order to meet the sales forecast of demand of:

January 60 units, February 70 units, March 80 units,
April 60 units, May 40 units, June 40 units

There is an opening stock of 50 units. The possibilities are that he/she puts a group of workers to work at a steady rate for the entire period or that he/she uses a larger group of workers and produces the units in batches. Present reasoned plans for both possibilities.

24 The managing director has been to a conference where he has heard of the great savings that can be made by using the just-in-time policy of inventory management. He has been concerned for some time with the capital tied up in stocks of raw materials and stocks of finished goods. At present, the purchasing department shops around for its raw materials and uses no particular supplier. Sometimes this leads to batches which have a high level of poor quality material but are cheap. The production department has a poor record regarding the quality of goods they produce and a large number of them are rejected as unacceptable when given a final inspection. However, this is felt to be the price to be paid for using a labour force which is paid at the lowest rates possible. There is little in the way of systematic training. Prepare a paper outlining (a) what is meant by just-in-time, and (b) the consequences of the company adopting such a policy.

25 The production manager of the XYZ Manufacturing Company has a problem in planning his/her budget for the forthcoming year. The sales department has presented their forecasts of demand for product X as:

Jan. 40 units, Feb. 50 units, Mar. 60 units, Apr. 100 units,
May 140 units, Jun. 120 units, Jul. 60 units, Aug. 40 units,
Sept. 20 units, Oct. 10 units, Nov. 10 units, Dec. 10 units.

The selling price per unit is £80. The production department working at full capacity can produce 200 units per month. The setting up cost per batch is £200, the holding cost per year per item is £10. At the beginning of January there is an initial stock of 10 units.

(a) Produce a plan for the production of this product. State any assumptions you make.
(b) Indicate how the stock of the product will vary from month to month.

8 Forecasting

8.1 Uses of forecasting

In businesses it is necessary to be able to make predictions about the future in order to plan the business operations. It is necessary to make *forecasts*. Thus, for example, a business might need to make forecasts about what products it should make in five years' time or more in order to plan for the introduction of new plant and possibly new factories. A production manager will need to be able to forecast what materials will be required for the next year so that production planning can occur.

Long-term forecasting is used by businesses to plan for such changes as the mix of products they offer, the development of new products, changes in production capacity, etc. Shorter-term forecasts will be used to determine information on which the requirements for personnel, equipment, and materials for the next year can be based. Sales forecasting is the basis of all financial budgets of departments in a company (see chapter 6). Thus the production budget which is concerned with the costs associated with the production of the products of the company is based on the sales forecasts of what products the company will produce and the number of items of each product required. The direct materials budget of what materials are required for production and the materials purchase budget of what materials should be purchased and their costs will then stem from the production budget. Predicting materials requirements is an essential ingredient of production control, since a lack of materials will result in production hold-ups and an excess of materials will result in unnecessary cost (see chapter 7 on inventory control).

The following are some of the uses of forecasting in organization function areas:

Short-term, i.e. a few months

1 Sales: sales force targets, sales by customer, sales by geographical area, sales of each type of product.
2 Production: demand for each type of product, plant loading, scheduling work.
3 Stores: demand for finished products, demand for materials, stock levels.
4 Finance: sales revenue, production costs, cash flow.
5 Purchasing: purchasing of materials taking into account lead times and demand.

Medium-term, i.e. a few months to a year

1 Sales: preparation of budgets, sales of products, new product introduction, range of products.
2 Production: preparation of budgets, introduction of new machinery and labour.
3 Stores: the demand for materials, inventory levels.

4 Finance: preparation of budgets, cash flow.
5 Purchasing: preparation of budgets, demand for materials.

Long-term, i.e. more than a few years
1 Sales: new product introduction
2 Production: expansion of plant, new production lines, ordering of major pieces of equipment
3 Stores: expansion of storage facilities
4 Finance: investment strategy, capital expenditure allocations.
5 Purchasing: long-term contracts for materials.

In addition, long-term forecasting is used by senior management to consider whether the objectives and strategies of the organization should be changed.

8.1.1 Forecasting techniques

This chapter offers an introduction to forecasting techniques. Forecasting techniques can be considered to fall into three categories, these being based on:

1 Opinion gathering.
2 Analysis of historical data.
3 Correlation with other variables that can be linked to product demand.

The gathering of opinions about what products will be required can by used for both medium and long-term forecasts. The analysis of historical data and correlation with other variables are techniques that are generally used for the short to medium term forecasts. Thus the question about what range of products a company should offer during the next few years might well be tackled by the gathering of opinions. The question of how large a stock of materials there should be during the next month and year might be tackled by the analysis of historical data. These techniques are looked at in more details in the following parts of this chapter.

8.2 Opinion gathering

Opinions about products and market trends can be gathered from a variety of sources. These might be technical staff and sales representatives within the company, product distributors, customers, or the general public. Within the company the technical staff and the sales representatives can provide information about the market for the products, though it has to be recognized that there may be some bias in respect to their opinions. Analysis of the reports of sales representatives from their meetings with customers can, however, lead to useful information about the products of the company. Product distributors, because they may well be selling products from a range of companies, may give more unbiased opinions. The collection of opinions from the general public and customers is generally based on sampling techniques. A sample of the general public or customers is carefully chosen to be representative of the public at large. The opinions of that sample are then taken, with some degree of uncertainty, to be representative of those of all the general public/customers. The opinions of the sample can be gathered by a number of methods, such as postal surveys, telephone surveys, field surveys involving personal interviews, product testing panels, etc.

8.3 Historical data analysis

To illustrate what is meant by historical data analysis, consider the situation where in January a company sold 20 units of product X, in February it sold 25 units, in March it sold 21 units, in April it sold 22 units, in May it sold 25 units, in June it sold 26 units. How many units might it envisage selling in July? The forecast for July can be based on considering these previous months' data, i.e. the historical data, and then using some technique to arrive at a figure for the forthcoming year. What we need to find out is the general trend in the data and so what might be expected for July.

We can plot the data on a graph and then see whether we can find a trend by establishing the best line to fit the points. The line can then be continued beyond the historical data to give the forecast for the future. Figure 8.1 illustrates this for the following data for the annual sales of the product given above. The dashed line on the graph would appear to have the data points reasonably equally scattered about it and so give the general trend of the data. Extrapolating the line beyond the points suggests that the forecast for July be 27 units.

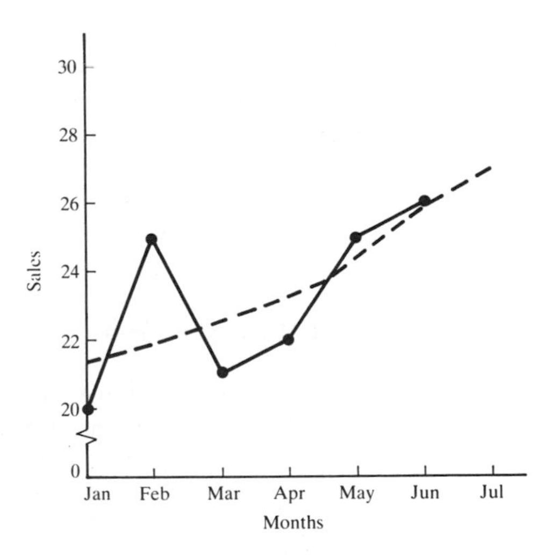

Figure 8.1 Trend graph

Example

By drawing the trend line based on the data given below, forecast the demand for product X in the month of June.

Month	Jan.	Feb.	Mar.	Apr.	May
Number of sales	100	155	140	150	180

Figure 8.2 shows the above data points plotted on a graph. They would appear to be scattered either side of a straight line. The straight line when extrapolated to June indicates a demand of 190 items.

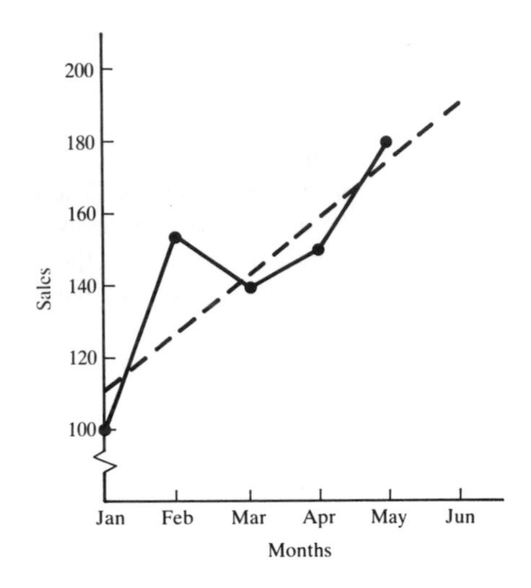

Figure 8.2 Example

8.3.1 Moving averages

A simple method of forecasting on the basis of historical data is the *moving average method.* With the graphical method discussed above, the trend line was estimated as being the line which smoothed out the fluctuations in the data to give the trend and so, on extrapolation, a forecast. The moving average method is a method of smoothing out data so that the trend line can be established.

The average of a set of n data points is the sum of that data divided by n, i.e.

$$\text{average} = \frac{\text{sum of all the data}}{n}$$

Taking the average of a number of points, determines the mean value about which the points fluctuate. The term *moving average* is used when each time the average is taken it is for the latest n set of data. Thus when the average is taken over, say, four pieces of data then we first take an average of the 1st to the 4th pieces of data. Then the next average is obtained by moving forward for the 2nd to 5th pieces of data. The next average is then taken by moving forward for the 3rd to 6th pieces of data. The set of data considered for the average moves forward each time by the dropping of the earliest piece of data and incorporating the latest piece. The larger the number of pieces of data over which the average is taken, the more the average irons out data fluctuations.

To illustrate the taking of moving averages, consider the sales figures for some product for the months January to July. The problem is to forecast the sales that might occur in August.

Jan.	Feb.	Mar.	Apr.	May	Jun.	Jul.	Aug.
8	10	9	13	12	14	13	

Suppose we take a moving average of the data for four months of data. For the first four months of January to April the average value is

$$\text{average} = \frac{8+10+9+13}{4} = 10.$$

The smoothed value for that period is thus 10. We can consider this smoothed value to be the value at the mid point of the period.

Jan.	Feb.	Mar.	Apr.	May	Jun.	Jul.	Aug.
[8	10	9	13]	12	14	13	

Average = 40/4 = 10

Now moving the data considered forward a month. The four month period now considered is February to May. The average for February to May is

$$\text{average} = \frac{10+9+13}{4} = 11$$

The smoothed value for the mid-point of that period is thus 11.

Jan.	Feb.	Mar.	Apr.	May	Jun.	Jul.	Aug.
8	[10	9	13	12]	14	13	

Average = 44/4 = 11

Now consider the data moved forward a month. The four month period now considered is March to June. This gives an average of 12 and so the smoothed value at the mid-point of the period is 12.

Jan.	Feb.	Mar.	Apr.	May	Jun.	Jul.	Aug.
8	10	9	13	12	14	13	

Average = 48/4 = 12

Now consider the data moved forward a month. The four month period now considered is April to July. This gives an average of 13 and so the smoothed value at the mid-point of the period is 13.

Jan.	Feb.	Mar.	Apr.	May	Jun.	Jul.	Aug.
8	10	9	13	12	14	13	

Average = 52/4 = 13

Figure 8.3 shows the data and the smoothed values plotted on a graph. Extrapolating the line through the smoothed points gives for August a sales forecast of 16 items.

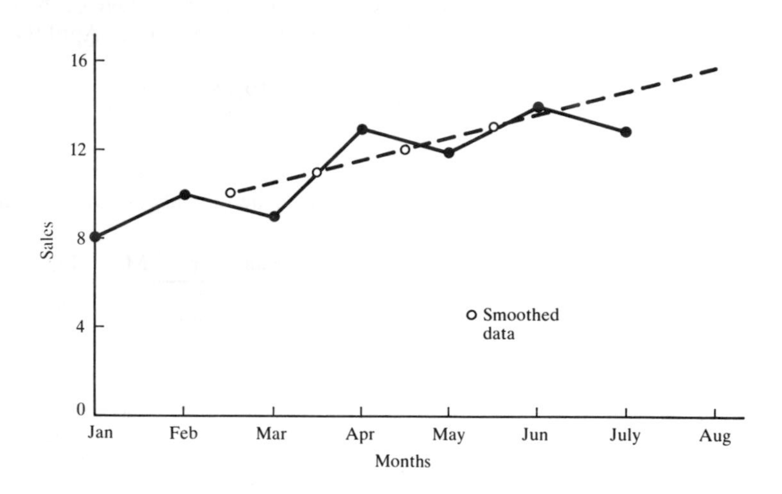

Figure 8.3 Moving average example

Example

The following data gives the sales figures for numbers of items of a product sold per month for the first six months of a year. Use three month moving averages to determine the trend and hence obtain a forecast for the month of July.

Jan.	Feb.	Mar.	Apr.	May	Jun.	Jul.
24	26	27	24	29	32	

Taking moving averages for each successive period of three months and putting the smoothed value at the mid-points of the three-month periods gives the following results:

Jan.	Feb.	Mar.	Apr.	May	Jun.	Jul.
24	26	27	24	29	32	
	25.6					
		25.6				
			26.6			
				28.3		

Figure 8.4 is a graph of the data and the trend line given by the moving average points. If the trend continues then we might expect sales of 32 items in July.

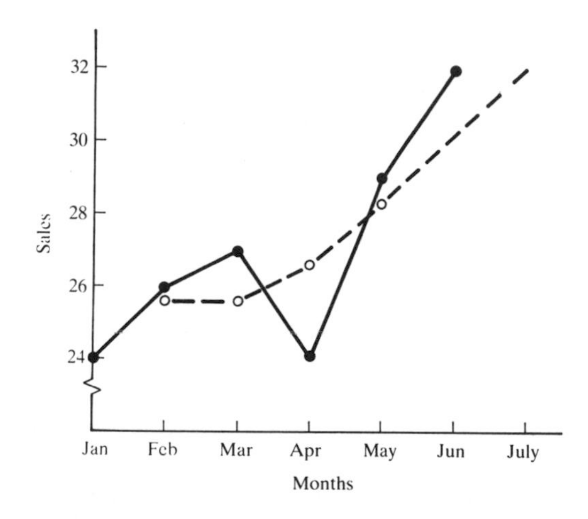

Figure 8.4 Example

8.3.2 Weighted moving averages

One of the problems of using moving averages is that the method gives equal weight to all the data. Thus, for example, for a three month period of January, February and March, the average gives equal weight to the data for each month. The value for January is treated exactly the same as that for February and that for March. Generally, however, the most recent period in a moving average will be the most significant. Thus the data for March is likely to be more significant in determining the trend for April than the data for January or February. The *weighted moving average* method is a method that can be used to give more weight to more recent data.

With a moving average over, say, four months we have

$$\text{average} = \frac{\text{sum of the four months}}{4}$$

$$= \tfrac{1}{4}\text{ month 1} + \tfrac{1}{4}\text{ month 2} + \tfrac{1}{4}\text{ month 3} + \tfrac{1}{4}\text{ month 4}$$

Each month is given the same weighting factor of ¼, the sum of the weighting factors being 1, i.e. ¼ + ¼ + ¼ + ¼ = 1. With a weighted moving average we can adjust the weighting to emphasize the more recent months, though we must still have the sum of the weighting factors equal to 1. Thus we might have

$$\text{weighted average} = \tfrac{1}{10}\text{ month 1} + \tfrac{2}{10}\text{ month 2} + \tfrac{3}{10}\text{ month 3} + \tfrac{4}{10}\text{ month 4}$$

Note that we have for the sum of the weighting factors

$$\tfrac{1}{10} + \tfrac{2}{10} + \tfrac{3}{10} + \tfrac{4}{10} = 1$$

To illustrate this method of weighted moving averages, consider the data used in the previous section with equally weighted moving averages for the sales of a product, i.e.

Jan.	Feb.	Mar.	Apr.	May	Jun.	Jul.	Aug.
8	10	9	13	12	14	13	

Taking a four month period with weighting factors of 1/10, 2/20, 3/10 and 4/10, then we obtain for the first four month period of January to April

$$\text{weighted average} = \tfrac{1}{10} \times 8 + \tfrac{2}{10} \times 10 + \tfrac{3}{10} \times 9 + \tfrac{4}{10} \times 13 = 10.7$$

For the next four month period of February to May, when the average moves forward one month, we have

$$\text{weighted average} = \tfrac{1}{10} \times 10 + \tfrac{2}{10} \times 9 + \tfrac{3}{10} \times 13 + \tfrac{4}{10} \times 12 = 11.5$$

For the next four month period of March to June

$$\text{weighted average} = \tfrac{1}{10} \times 9 + \tfrac{2}{10} \times 13 + \tfrac{3}{10} \times 12 + \tfrac{4}{10} \times 14 = 12.3$$

For the next four month period of April to July

$$\text{weighted average} = \tfrac{1}{10} \times 13 + \tfrac{2}{10} \times 12 + \tfrac{3}{10} \times 14 + \tfrac{4}{10} \times 13 = 13.1$$

Thus, if we put the weighted average as the smoothed value at the centre of each of the four month periods:

Jan.	Feb.	Mar.	Apr.	May	Jun.	Jul.	Aug.
8	10	9	13	12	14	13	
	10.7						
		11.5					
			12.3				
				13.1			

Figure 8.5 shows the graph of the weighted average points and the resulting trend line drawn through them. With this trend, the forecast for August is 15 items. Compare this result with the forecast given earlier, with the same data, for the unweighted moving averages (figure 8.3) of 16. The result is a smaller forecast because more weight has been given to the sales for the latest month of July. It is less influenced by the high figure for June and more influenced by the lower figure of 13 for July.

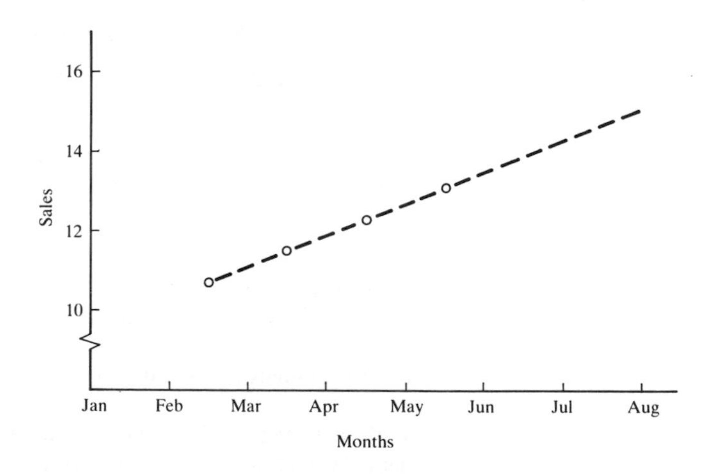

Figure 8.5 Weighted averages example

Example

The following data gives the sales figures for numbers of items of a product sold per month. Use three month weighted moving averages, with weights 1/6, 2/6, 3/6, to determine the trend and hence a forecast for the month of July. Note this is the example considered in the previous section with unweighted moving averages.

Jan.	Feb.	Mar.	Apr.	May	Jun.	Jul.
24	26	27	24	29	32	

Considering the months January to March then the weighted average is

$$\text{weighted average} = \tfrac{1}{6} \times 24 + \tfrac{2}{6} \times 26 + \tfrac{3}{6} \times 27 = 26.2$$

For the months February to April,

$$\text{weighted average} = \tfrac{1}{6} \times 26 + \tfrac{2}{6} \times 27 + \tfrac{3}{6} \times 24 = 25.3$$

For the months March to May,

$$\text{weighted average} = \tfrac{1}{6} \times 27 + \tfrac{2}{6} \times 24 + \tfrac{3}{6} \times 29 = 27.0$$

For the months April to June,

$$\text{weighted average} = \tfrac{1}{6} \times 24 + \tfrac{2}{6} \times 29 + \tfrac{3}{6} \times 32 = 29.7$$

Thus we have:

Jan.	Feb.	Mar.	Apr.	May	Jun.	Jul.
24	26	27	24	29	32	
	26.2					
		25.3				
			27.0			
				29.7		

Figure 8.6 shows the results of the smoothing operation. Extrapolation to July indicates a sales figure for July of 34.5 items.

8.3.3 Exponential moving averages

With weighted moving averages, more weight is given to the most recent data. Weightings for a four period moving average might be, for most recent to most distance periods, 4/10, 3/10, 2/10, 1/10. Thus the weightings are decreased by a constant factor of 1/10 in moving period by period from the most recent to the most distant of the data for the four periods. A problem with this method is that it is rather arbitrary as to how many periods we choose to consider.

Another possibility for the weighting is to decrease the weighting by a constant factor in moving period by period from the most recent to the most distant data. With this method we do not have to specify the number of periods involved. For example, if we have a weighting figure which is halved in moving from one period to the next then we might have weighting factors of 1/2, 1/4, 1/8, 1/16, 1/32, 1/64, 1/128, ... There is no cut off to the number of periods, the weighting factors just get smaller and smaller. This method of smoothing is known as *exponential moving averages* or *exponential smoothing*.

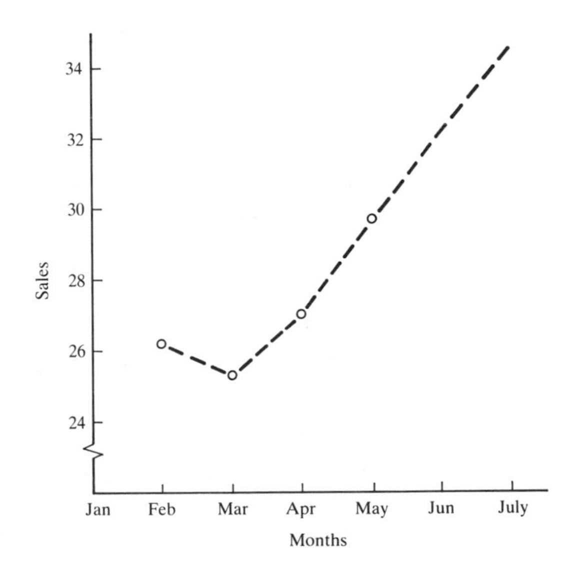

Figure 8.6 Example

We can write the series of numbers 1/2, 1/4, 1/8, 1/16, etc. as

$$\tfrac{1}{2},\ \tfrac{1}{2}(1-\tfrac{1}{2}),\ \tfrac{1}{2}(1-\tfrac{1}{2})(1-\tfrac{1}{2}),\ \tfrac{1}{2}(1-\tfrac{1}{2})(1-\tfrac{1}{2})(1-\tfrac{1}{2}),\ \ldots$$

Thus, in general, the weighting factors used are written as

$$\alpha,\ \alpha(1-\alpha),\ \alpha(1-\alpha)(1-\alpha),\ \alpha(1-\alpha)(1-\alpha)(1-\alpha),\ \ldots$$

where α is a constant called the *smoothing constant*. It has a value between 0 and 1. Thus with $\alpha = 0.2$ we have weighting factors of

$$0.20,\ 0.16,\ 0.128,\ 0.1024,\ 0.08192,\ 0.065536,\ \ldots$$

The series of weighting factors has a sum of 1 when summed to infinity. Using such weighting factors we could determine the smoothed value at a particular time by summing all the weighted values due to all the times preceding. However, there is a more useful way of using such a method.

Suppose we calculate a smoothed value at a particular time by summing all the terms from that time backwards in time. Then when we move forward to the next interval of time we could repeat the calculation taking into account the newer period. However, it can be shown that

smoothed value at time t
= $\alpha \times$ data value at time t
+ $(1 - \alpha) \times$ smoothed value obtained for time $(t - 1)$

The following shows how the above equation can be derived. The smoothed value F_t for a time t is given by

$$F_t = \alpha x_t + \alpha(1-\alpha)x_{t-1} + \alpha(1-\alpha)(1-\alpha)x_{t-2} + \ldots$$

where x_t is the data value at time t, x_{t-1} is the data value at time $(t - 1)$, x_{t-2} is the data value at time $(t - 2)$, etc. We can write this equation as

$$F_t = \alpha x_t + (1-\alpha)[\alpha x_{t-1} + \alpha(1-\alpha)x_{t-2} + \ldots]$$

But the term in the square brackets is just the smoothed F_{t-1} that would have been obtained for the time $(t - 1)$. Hence we can write

$$F_t = \alpha x_t + (1-\alpha)F_{t-1}$$

This is the equation that was given earlier.

The equation is often found written in a different form to make calculations easier. Rearranging the terms gives

$$F_t = \alpha x_t + F_{t-1} - \alpha F_{t-1}$$

$$= F_{t-1} + \alpha(x_t - F_{t-1})$$

Thus we have

smoothed value at time t
= smoothed value at time $(t - 1)$
+ α[value at time t – smoothed value at time $(t - 1)$]

The smoothed value obtained at a time t is often taken as the forecast for the time $(t + 1)$. Likewise the smoothed value obtained at time $(t - 1)$ is taken as the forecast for time t. This is reasonable if the trend indicates only slowly changing data. With this, we then have

forecast for time $(t + 1)$ = forecast for time $(t - 1)$
+ α[value at time t – forecast for time t]

The following example illustrates the use of exponential smoothing. Consider the data used earlier to illustrate the moving averages and weighted moving averages methods for sales of a particular product.

Jan.	Feb.	Mar.	Apr.	May	Jun.	Jul.	Aug.
8	10	9	13	12	14	13	

We will use, for this example, a smoothing factor of 0.2. It is common practice to take the first value as being the first smoothed value. Thus the smoothed value in January is taken as 8. Using the equation from above,

smoothed value at time t
= smoothed value at time $(t - 1)$
+ α[value at time t – smoothed value at time $(t - 1)$]

then for February we have

Feb. smoothed value = smoothed value in Jan.
+ α(Feb. value – smoothed value for Jan.)

$$= 8 + 0.2(10 - 8) = 8.4$$

Hence we have

	Jan.	Feb.	Mar.	Apr.	May	Jun.	Jul.	Aug.
	8	10	9	13	12	14	13	
Smoothed value		8.4						

For March we have

Mar. smoothed value = smoothed value in Feb.
+ α(Mar. value – smoothed value for Feb.)

$$= 8.4 + 0.2(9 - 8.4) = 8.52$$

and so we now have

	Jan.	Feb.	Mar.	Apr.	May	Jun.	Jul.	Aug.
	8	10	9	13	12	14	13	
Smoothed values		8.4	8.52					

For April we have

Apr. smoothed value = smoothed value in Mar.
+ α(Apr. value – smoothed value for Mar.)

$$= 8.52 + 0.2(13 - 8.52) = 8.83$$

and so we now have

	Jan.	Feb.	Mar.	Apr.	May	Jun.	Jul.	Aug.
	8	10	9	13	12	14	13	
Smoothed values		8.4	8.52	8.83				

For May we have

$$\text{May smoothed value} = \text{smoothed value in Apr.} + \alpha(\text{May value} - \text{smoothed value for Apr.})$$

$$= 8.83 + 0.2(12 - 8.83) = 9.46$$

	Jan.	Feb.	Mar.	Apr.	May	Jun.	Jul.	Aug.
	8	10	9	13	12	14	13	
Smoothed values		8.4	8.52	8.83	9.46			

For June we have

$$\text{June smoothed value} = \text{smoothed value in May} + \alpha(\text{June value} - \text{smoothed value for May})$$

$$= 9.46 + 0.2(14 - 9.46) = 10.37$$

	Jan.	Feb.	Mar.	Apr.	May	Jun.	Jul.	Aug.
	8	10	9	13	12	14	13	
Smoothed values		8.4	8.52	8.83	9.46	10.37		

For July we have

$$\text{July smoothed value} = \text{smoothed value in June} + \alpha(\text{July value} - \text{smoothed value for June})$$

$$= 10.37 + 0.2(13 - 10.37) = 10.90$$

and thus the smoothed data is

	Jan.	Feb.	Mar.	Apr.	May	Jun.	Jul.	Aug.
	8	10	9	13	12	14	13	
Smoothed values		8.4	8.52	8.83	9.46	10.37	10.90	

Figure 8.7 shows the exponentially smoothed data plotted on a graph. The forecast for August might then be 11 items.

The smaller the value taken for the smoothing constant the smoother will be the resulting trend graph. This is because the smaller the value of the constant the more significant become the preceding terms in determining the smoothed value. Effectively, more terms have an effect on the value obtained and so the less the effect of any one term markedly deviating from the norm. If the above calculation is repeated with smoothing factors of 0.5 and 0.8 then the results are:

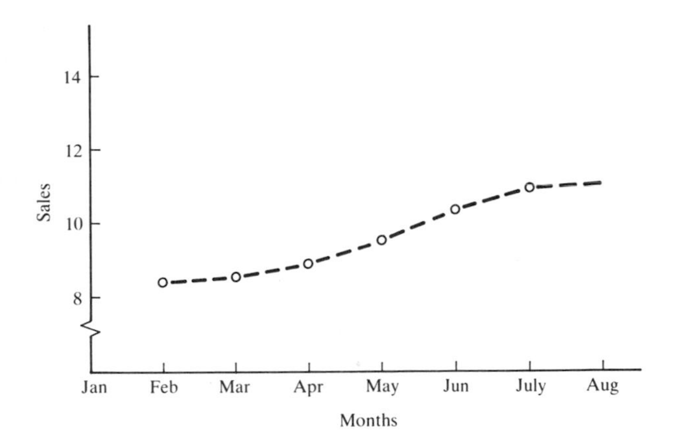

Figure 8.7 Exponential smoothing example

	Jan.	Feb.	Mar.	Apr.	May	Jun.	Jul.	Aug.
	8	10	9	13	12	14	13	
Smoothed values $\alpha = 0.2$		8.4	8.52	8.83	9.46	10.37	10.9	
Smoothed values $\alpha = 0.5$		9	9	11	11.5	12.75	12.89	
Smoothed values $\alpha = 0.8$		9.6	8.46	12.09	12.09	13.31	13.06	

Note that the higher the smoothing factor the greater the effect of the July sales figure on the smoothed value obtained.

Example

The following data gives the sales figures for numbers of items of a product sold per month. Use exponential smoothing with a smoothing factor of 0.6 to determine the trend and hence a forecast for the month of July. Note this is the example considered in the previous sections with weighted moving averages and unweighted moving averages.

Jan.	Feb.	Mar.	Apr.	May	Jun.	Jul.
24	26	27	24	29	32	

Taking the first value as being the first smoothed value then for February we have

$$\text{Feb. smoothed value} = \text{smoothed value in Jan.} + \alpha(\text{Feb. value} - \text{smoothed value for Jan.})$$

$$= 24 + 0.6(26 - 24) = 25.2$$

For March we have

$$\text{Mar. smoothed value} = \text{smoothed value in Feb.} + \alpha(\text{Mar. value} - \text{smoothed value for Feb.})$$

$$= 25.2 + 0.6(27 - 25.2) = 26.3$$

For April we have

$$\text{Apr. smoothed value} = \text{smoothed value in Mar.} + \alpha(\text{Apr. value} - \text{smoothed value for Feb.})$$

$$= 26.3 + 0.6(24 - 26.3) = 24.9$$

For May we have

$$\text{May smoothed value} = \text{smoothed value in Apr.} + \alpha(\text{May value} - \text{smoothed value for Apr.})$$

$$= 26.3 + 0.6(29 - 26.3) = 28.5$$

For June we have

$$\text{June smoothed value} = \text{smoothed value in May.} + \alpha(\text{Jun. value} - \text{smoothed value for May})$$

$$= 28.5 + 0.6(32 - 28.5) = 30.6$$

Thus we have:

	Jan.	Feb.	Mar.	Apr.	May	Jun.	Jul.
	24	26	27	24	29	32	
Smoothed values		25.2	26.3	24.9	28.5	30.6	

If we assume that the July forecast is the June smoothed value then the forecast would be 30.6 items. However, if we plot the smoothed values on a graph (figure 8.8) and use the trend line then extrapolation of the smoothed data suggests a forecast of 33 items for the month of July.

A smaller smoothing constant would have given a smoother graph with the odd April figure then having less effect on the trend line.

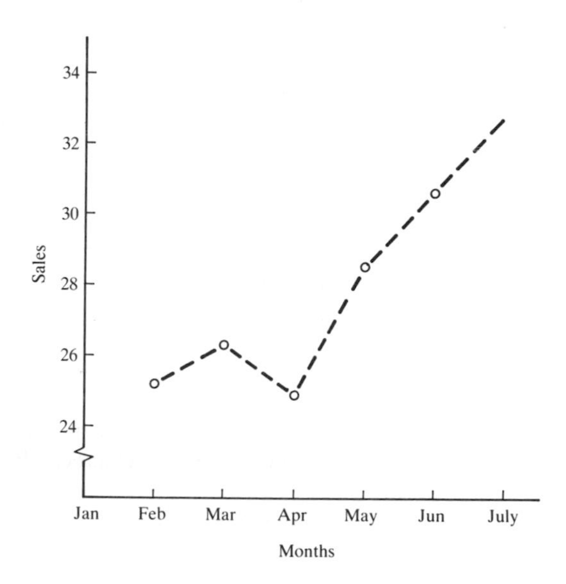

Figure 8.8 Example

Correlation with related variables

It is sometimes possible to make forecasts about the value of one quantity if there is information about the value of a different but related quantity. The two variables are said to be *correlated* if they are linked in this way. For example, the number of items of a particular product that are sold per month might correlate to the selling price of the product. When the sale price is increased then the number of items sold decreases. The prices of goods might correlate with the prices of raw materials, the higher the price of raw materials the higher the selling price of the goods. The sales of raw materials, e.g. steel bar, might correlate with the general level of industrial production. When the level of production rises then the demand for steel bar rises.

In some cases there may be coincidence in the timing of the changes in the two related variables, in some cases a change in one variable may lag or lead the change in the other variable. Thus the change in the number of items sold might occur at the same time as the change in the selling price. The prices of goods might lag behind the prices of raw materials.

Example

Production of a particular product is in small batches with sizes dictated by the orders received. A production manager wishes to establish a relationship between the direct labour costs for the product and the size of the batch. The following is information that he/she has gathered on previous batches. Determine whether there is a simple relationship between direct labour costs and batch size.

Batch size	15	18	20	24	27
Direct labour costs £	200	240	270	320	380

Figure 8.9 shows the data plotted as a graph. There appears to be a straight line which can be drawn through the points. This straight line has the slope of about 150/10 = £15 per batch and would intercept the costs axis at 0 when the batch size is 0. The equation of the straight line is thus

$$\text{Cost} = 15 \times \text{batch size}$$

Thus, for example, if the production manager was required to produce a batch of 22 items, the direct labour cost would be

$$\text{cost} = 15 \times 22 = £330$$

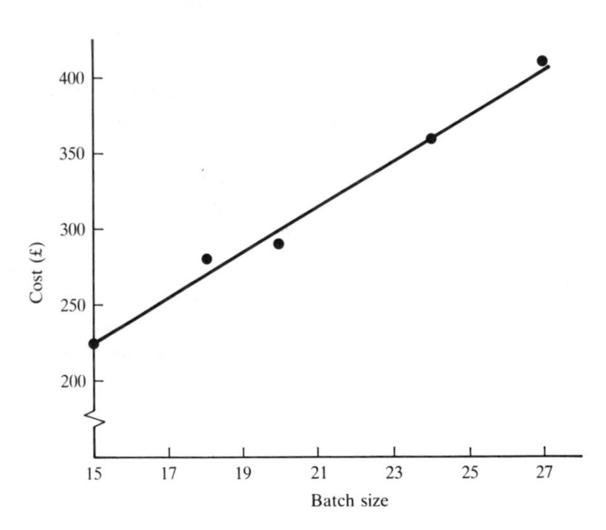

Figure 8.9 Example

Problems

Review questions

1 Give examples of the uses that can be made of short, intermediate and long-term forecasts in (a) sales, (b) production.

2 Explain how moving averages can be used to smooth out fluctuations in data and give a trend line.

3 Explain the reasons for using weighted moving averages rather than just simple moving averages to smooth out fluctuations and give a trend line.

4 Explain what is meant by exponential smoothing.

5 If the sales for 19X1 were 80 items of a particular product, for 19X2 75 items, and for 19X3 82 items, determine the average value of the sales during those three years.

6 If the sales for 19X1 were 80 items of a particular product, for 19X2 75 items, and for 19X3 82 items, determine the weighted average value of the sales during those three years if the weighting factors are 1/6, 2/6, 3/6.

7 If the sales for 19X1 were 80 items of a particular product, for 19X2 75 items, for 19X3 82 items, for 19X4 78 items, determine the smoothed values for 19X2, 19X3 and 19X4 using exponential smoothing with a smoothing constant of (a) 0.2, (b) 0.9.

8 Forecasts for the sales of machine tools by a company are made by the use of the equation

$$\text{forecast sales} = 0.30(\text{last year's sales}) + 0.70(\text{forecast of last year's sales}$$

What is the value of the smoothing factor being used in this exponential smoothing method of forecasting?

9 The sales of machine tools are found to be related to the bank interest rate prevailing at the time. If the relationship is

$$\text{number of sales} = \frac{1000}{\text{bank interest rate}}$$

What will be the forecast sales when the bank interest rate is 5%?

Multiple choice questions
For problems 10 to 18, select from the answer options A, B, C or D the one correct answer.

10 Decide whether each of these statements is True (T) or False (F).

(i) A moving average is an average where the number of periods over which the average is taken changes.
(ii) A weighted moving average can be used to give most weight to the most recent data.

Which option BEST describes the two statements?
A (i) T (ii) T
B (i) T, (ii) F
C (i) F (ii) T
D (i) F (ii) F

Problems 11-13 relate to the following information:

The sales figures for a product are:

Jan.	Feb.	Mar.	Apr.
12	17	13	18

11 The unweighted moving averages for the sales data for three month periods are:

A 1st 3 month period: 13, 2nd 3 month period: 18
B 1st 3 month period: 14, 2nd 3 month period: 16
C 1st 3 month period: 14, 2nd 3 month period: 15
D 1st 3 month period: 17, 2nd 3 month period: 13

12 The weighted moving averages for the sales data for three month periods with weightings of 1/6, 2/6, 3/6 for the distant to recent data points are:

A 1st 3 month period: 7.0, 2nd 3 month period: 5.7
B 1st 3 month period: 14.0, 2nd 3 month period: 16.0
C 1st 3 month period: 14.2, 2nd 3 month period: 16.2
D 1st 3 month period: 17.0, 2nd 3 month period: 13.0

13 The exponentially weighted averages, using a smoothing constant of 0.5, are:

A For February 14.5, for March 15.75
B For February 15.0, for March 19.0
C For February 20.5, for March 23.5
D For February 23.0, for March 21.5

14 Decide whether each of these statements is True (T) or False (F).

With the exponential smoothing method of determining the trend among data and so making forecasts:

(i) A large smoothing constant means that more weight is attached to the latest data that is the case with a small smoothing constant.
(ii) The average is only taken over just the last two data points and is only affected by their values.

Which option BEST describes the two statements?

A (i) T (ii) T
B (i) T, (ii) F
C (i) F (ii) T
D (i) F (ii) F

Problems 15-17 relate to the following information.

The amount of steel bar drawn from stock in a company in the four quarters of two successive years are:

Year 1
1st quarter 710 units
2nd quarter 1400 units
3rd quarter 970 units
4th quarter 800 units

Year 2
1st quarter 780 units
2nd quarter 1500 units
3rd quarter 1020 units
4th quarter 852 units

15 Using four-period moving averages, the smoothed value for the last four periods is:

A 852
B 970
C 1004
D 1038

16 Using a weighted four-point moving average, with weightings of 1/10, 2/10, 3/10 and 4/10, the smoothed value for the first four periods is:

A 320
B 962
C 970
D 1004

17 Using exponential smoothing with a smoothing constant of 0.2, the smoothed value for the second period of the first year is:

A 142
B 280
C 848
D 1512

18 A salesman considers that the sales of icecream are related to the temperature at noon on a day by the relationship

$$\text{sales in £} = 10 \times (\text{temperature in °C}) + 100$$

The sales on a day when the temperature at noon is 22°C are thus forecast to be:

A £100
B £120
C £220
D £320

Assignments

19 Obtain data for sales of some item from a store, e.g. a college stationery store, and use them to develop a forecasting method for future sales of that item.

20 Obtain data concerning the pass rates on the course your are studying and use them to develop a forecasting method for future pass rates. What is the pass rate forecast for your year?

Case studies

21 The ABC Electronics Company is a small company which has only been in existence for a few years. Initially it was extremely profitable, however it is now having financial problems. The managing director feels that this is due to the large amount of money tied up in stocks of in some finished goods and the problems occurring with other goods due to a lack of finished goods which would enable them to meet customers' orders. At present the only forecasting method used is the assumption that next year will always be better than the previous year and so production is geared up at the beginning of the year to produce 10% more of each of the products.

(a) Suggest a forecasting method that might be used.
(b) Use your forecasting method to develop a forecast for the year 19X6 for the product which in previous years had the following sales:

19X1	19X2	19X3	19X4	19X5
400	440	480	530	600

How accurate would your method have been in forecasting sales for the year 19X5 given the previous years' data?

22 The owners of a market stall which sells cups of coffee are having a problem working out how many plastic cups and how much coffee to purchase and take along to the market on market days. If they take too little they lose sales, if they take too much they have tied up their capital in stock that is not used on the day. They have gathered the data on the number of plastic cups used for a number of weeks and obtained the following:

Week 1	Week 2	Week 3	Week 4	Week 5
230	256	190	255	290

(a) Devise a forecasting model that they might use.
(b) On the basis of your model, forecast the number of cups they are likely to need in week 6. Hence, allowing some buffer stock, suggest how many cups they should take.
(c) In forecasting the amount of coffee to take, a model that might be used is to expect some correlation with the number of cups used. If each cup of coffee requires 10 g of coffee, forecast the amount of coffee likely to be needed in week 6.

Appendix: Assignments

Assignment-based learning

This topic of Engineering and Commercial Functions of Business lends itself to an assignment-based form of learning, involving core skills and possibly integrated with assignment work in other units. An assignment involving a manufacturing business could be used for the basis of all the learning in this topic. This could be a hypothetical company with all details being worked out on paper or a computer screen. Alternatively, students could work together as a "real" business and actually make and sell items. This obviously would involve integrating this topic with other topics such as Engineering Processes, Engineering Materials, Design Development, etc.

Example

The following example illustrates how the parts of the assignment might be developed for a hypothetical business when it is a paper or computer screen exercise. By the end of this topic, the details of the company and how it functions should have been worked out.

1 Establish the basic parameters of your hypothetical company.

   Consider, for example, a small manufacturing company which makes just a single product. Decide initially on some product which your hypothetical company will be making. Keep the product simple, with few component parts, so that the tasks involved in determining costs will not become too onerous. You might like to consider for the product one such as a mains plug, a bicycle pump, a pocket size radio. The product chosen should be one where it is possible to comprehend how it might be manufactured, the machines that might be involved in a small company, and the materials and components used.

2 Specify the types of functional activities might be expected to occur in the company. (Chapter 1)

3 Specify what type of organization it is, i.e. primary, secondary or tertiary. (Chapter 1)

4 What links will the organization might it need with other organizations, i.e. establish a feasible chain of production? (Chapter 1)

5 Consider what types of roles engineers might have in the organization. (Chapter 1)

6 Decide on a feasible labour force for the company and an organizational structure for it. Take into account the need for specialization and the span of control that might be feasible for managers/supervisors, also the type of production operations to be used, i.e.. one-off, batch, flow or

process. Consider whether the structure might be functional and work process, by location or matrix. (Chapter 2)

7 Write job descriptions for a production worker, a production supervisor/manager and a senior manager. (Chapter 2/3)

8 Specify the activities of the various departments/functions in the organization. (Chapter 3)

9 Explain how quality of the products will be assured. (Chapter 4)

10 Determine how the company will determine what form its product should take and how many it should produce. (Chapter 4)

11 Specify the information flow pattern within the company for the development of a new product. (Chapter 4)

12 Specify the information flow pattern for the procedures involved from receiving an order to despatch of the order to the customer and payment by the customer. (Chapter 4)

13 Consider, in general, the costs involved in making the product, listing the fixed and the variable costs. Depending on the product, you might wish to consider the costs of alternative processes and/or materials. (Chapter 5)

14 Consider the selling price of the product and hence the volume of sales that would be needed for the company to make a profit. (Chapter 5)5 Consider how the direct materials, direct labour and overheads costs are to be worked out and work them out for your product. (Chapter 6)

16 Use absorption costing and marginal costing with regard to your product and company in order to determine how the costs and revenue work out. (Chapter 6)

17 Establish budgets for the various parts of your company and an overall master budget. (Chapter 7)

18 Explain how inventories are to be controlled in your company. Could you operate just-in-time? (Chapter 8)

19 Establish the methods to be used for forecasting sales and make forecasts. (Chapter 9)

Answers

The following indicate the sections of a chapter in which the answers to revision questions may be found and the answers to the multiple-choice questions.

Chapter 1

1 See section 1.1
2 See section 1.1.1
3 See section 1.2.1
4 See section 1.3
5 See section 1.3
6 See section 1.3.1
7 A
8 D
9 B
10 D
11 A
12 B
13 D
14 D
15 C
16 D

Chapter 2

1 See section 2.1
2 See section 2.3
3 See section 2.2
4 See section 2.3.1
5 See section 2.5
6 A
7 C
8 B
9 A
10 B
11 D
12 D
13 A
14 A
15 B
16 C
17 A

Chapter 3

1 (a) See section 3.1.4, items 1, 2, 3, 4, (b) See section 3.1.8, (c) see sections 3.1.5, items 1, 2, 3, 4, (d) see section 3.1.9, item 5.
2 See section 3.1.7
3 See figure 3.5

4 See section 3.5
5 Eg. (a) need for a new product, (b) production information, (c) budget for production, (d) details of stock ordered.
6 See section 3.2.1
7 B
8 C
9 A
10 B
11 C
12 D
13 B
14 C
15 D
16 D
17 D

Chapter 4

1 See section 4.1
2 £0.40
3 £10 000
4 £27.50
5 See section 4.5.1
6 1000
7 £22
8 As figure 4.8, methods B and A
9 (a) £4400, (b) £60 000, (c) Loss £4000, (d) Profit £16 000, (e) 5000
10 (a) £450 000, (b) £750 000, (c) Loss £30 000, (d) Profit £90 000, (e) 25 000
11 See section 4.6
12 A
13 D
14 C
15 B
16 A
17 C
18 B
19 B
20 B
21 A
22 C

Chapter 5

1 See section 5.1
2 See section 5.2
3 See section 5.6
4 See section 5.6.1
5 £8
6 X £1.50, Y £2.25
7 Administration £14 000, production £28 000, stores £8 000
8 X £10 000, Y £4 000, Z £6 000
9 See section 5.7

10 £4000
11 (a) £120, (b) £136, (c) £128
12 See section 5.1.1
13 A
14 C
15 C
16 B
17 B
18 B
19 D
20 D
21 B
22 A
23 B
24 D
25 C

Chapter 6

1 See section 6.1.1
2 See sections 6.2.1, 6.2.3, 6.2.5, 6.2.7
3 See section 6.3
4 See section 6.2.7
5 See section 6.5
6 See section 6.6
7 C
8 D
9 B
10 D
11 D
12 D
13 A
14 C
15 D
16 C
17 D
18 B
19 C

Chapter 7

1 See section 7.1.1
2 See section 7.2
3 174.6
4 63.2
5 3 weeks
6 200
7 A
8 A
9 C
10 A
11 D
12 B

13 B
14 C
15 C
16 A
17 B
18 C

Chapter 8

1 See section 8.1
2 See section 8.2.1
3 See section 8.2.2
4 See section 8.2.3
5 79
6 79.3
7 (a) 79.0, 79.6, 79.3, (b) 75.5, 81.4, 78.3
8 0.70
9 200
10 C
11 B
12 C
13 A
14 B
15 D
16 B
17 D
18 D

Index